Speaking of Science

Notable Quotes on Science, Engineering, and the Environment

Compiled
and Edited by
**Jon Fripp,
Michael Fripp, and
Deborah Fripp**

By necessity, by proclivity,
and by delight, we all quote.
— *Ralph Waldo Emerson*
in *Letters and Social Aims*

Technology Publishing
Eagle Rock, VA
www.LLH-Publishing.com

The wisdom of the wise and the experience of the ages
are perpetuated by quotations.
— *Benjamin Disraeli, British Prime Minister*

A proverb: A short saying that oft contains much wisdom.
—*Sophocles, 496-406 B.C.E.*

Library of Congress Cataloging-in-Publication Data
Speaking of science: notable quotes on science, engineering, and the
environment/compiled and edited by Jon Fripp, Michael Fripp, and
Deborah Fripp.
    p. cm.
  Includes bibliographical references and index.
  ISBN 1-878707-51-5 (pbk : alk. paper)
  1. Science—Quotations, maxims, etc. 2. Engineering—
Quotations, maxims, etc. 3. Scientists—Biography. I. Fripp, Jon, 1967–
II. Fripp, Michael, 1970– III. Fripp, Deborah, 1970–

Q173.S738   2000
500—dc21

                                              99-045212

Printed in the U.S.A.
10 9 8 7 6 5 4 3 2 1

Foreword

Welcome to the first edition of *Speaking of Science: Notable Quotes on Science, Engineering, and the Environment*. We hope that you enjoy it.

Quotations are powerful tools and are an important component of clear communication. They can be used to persuade, to instruct, and to illuminate deeper meanings. Quotes can serve as mnemonic devices in teaching and as focal points for reflection and insight. Quotes can bridge time and tie a modern idea to a time-honored meaning. Quotes also can entertain through irreverence or the juxtaposition of conflicting ideas.

We, the authors, use quotes in our public speaking, in our technical writing, in our teaching, and in our daily conversations. We like quotes and, thus, started collecting quotes. Other people liked our quote collection and asked for copies. This quote book was born from the obvious need for a repository of quotes for technical speakers. This book contains quotes from across the ages with speakers from the 20th century B.C.E. through the 20th century C.E.*

Traditionally, quotes and, more specifically, quote books have been the purview of poets and politicians. Such a narrow focus misses the rich sources of more technical subjects. This book seeks to fill the quotational void by tailoring the selection of quotes to the worldview of scientists, engineers, and environmentalists.

Technical quotes are interesting because they often represent snapshots of scientific progress. We have tried to include quotes voicing disparate points of view. In some cases, these quotes articulate opposing points of view while in other cases these quotes present the evolution of scientific thought.

Following many of the quotes, we have added information to explain the context of the quote. This "value added" addition often includes relevant information about the speaker. More information about the speakers is located at the end of the book. In some cases, we have included our own thoughts about the quote; we hope that you enjoy our musings.

Needless to say, the quotes have been derived from other publications. Wherever possible, credit has been given to the publication in which the quotes were originally published. We have been collecting quotes for longer than we have been contemplating publishing a book and, hence, some of the publication references are incomplete. We would appreciate feedback from anyone in the reading audience who has better reference information. That information will be included in subsequent editions of this book.

---

*B.C.E. stands for Before Common Era and C.E. stands for Common Era. These are more ecumenical notations for B.C. which stands for Before Christ and A.D. which stands for In the Year of Our Lord.

Acknowledgments

We have received significant assistance in the creation of this book. Foremost, we would like to thank all of the speakers who said the wonderful things that have been quoted here. Without them, this book would not exist.

The quote collection has been accumulated over many years. The authors have received assistance not only in the collection process but also in the formulation of this book from Daniel Fripp, Allison Fripp, Archie Fripp, Jean Fripp, Valeska Fripp, Joe Redish, Bill Smollen, Jill Caverly, Chris Spaur, Chuck Dietz, Craig Fischenich, Leslie Flanagan, Robert Bank, Ken Halstead, Shannon Bard, Eli Hestermann, John Enright, and Kristie Thayer, among others. Additionally, we appreciate the publication assistance provided by Carol Lewis at LLH Technology Publishing.

We also received assistance from many texts. The texts are listed in the section titled "References" located at the end of the book. The biographies of the quote speakers were created with help from Microsoft's Encyclopedia Encarta 98.

All of the cartoons were drawn in Corel Draw 8.0 with Corel clipart. The images of money featuring famous scientists were collected by Joe Redish. His complete collection of scientists on currency can be viewed at www2.physics.umd.edu/~redish/Money/

Contents

Foreword v
Acknowledgments vii

Science

Chapter 1 1
Physics 4
Weights and Measures 9
Materials Science 11
Chemistry 13
Biology 16
Animal Behavior 21
Evolution 23
Cloning 27
Archaeology 29
Anthropology 30
Research 31
Behavior of Scientists 36
Divisions in Science 39

Mathematics

Chapter 2 43
Proofs 47
Statistics 49
Theory and Modeling 52
Math in Science 56
Balance Between Theory and Reality 58
Relevance of Math 62
Math as Art 66
Mathematicians 67

Engineering

Chapter 3 75
Mechanical Engineering 77
Engineering Design 79
Aeronautical Engineering 81
Astronautical Engineering 84
Electrical Engineering 89
Atomic Power 92
Computer Science 94
Transportation 97
Civil Engineering 99
Hydraulic Engineering 104
Irrigation 107
Flood Control 109
Life of the Engineer and Scientist 112
Technological Development 116

Man and the Environment	**Chapter 4**	121
	Habitat Destruction	122
	Man's Impact on the Environment	124
	Travels on the Water	127
	Water Quality	129
	Power of Water	131
	Conservation	132
	Restoration	135
	Power of Nature	138
	Sustainable Development	140
Nature	**Chapter 5**	143
	Plants	147
	Forests	150
	Wetlands	151
	Rivers	152
	Oceans	157
	Whales and Dolphins	159
	Animals	163
	Biodiversity	166
Teaching Science	**Chapter 6**	169
	Technical Writing	170
	Presentations	174
	Students	175
	Teachers	177
	Teaching Technique	178
	Value of Education	181
	Universities	183
The Working Environment	**Chapter 7**	187
	Committees	187
	Planning and Implementation	190
	Human Behavior	193
	Communication	196
	Call to Action	200
	Trust	204
	Pretension	205
	Money and Greed	207
	Common Sense	209
	Odds and Ends	210
	References	213
	Biographies	217
	Index	233
	About the Editors	241

Chapter 1
Science

We especially need imagination in science. It is not all mathematics, nor all logic, but it is somewhat beauty and poetry.

> — *Maria Mitchell, 1866*
> Quoted in *Maria Mitchell, Life, Letters, and Journals*
> by Phebe Mitchell Kendall
> *The words of great minds can often serve to stimulate such imagination.*

Science is mind applied to nature.

> — *Alexander von Humboldt*
> In *Cosmos, a Sketch of the Universe*, 1848

The most beautiful thing we can experience is the mysterious. It is the source of all true art and science.

> — *Albert Einstein*
> In the lecture "What I Believe," 1930

Albert Einstein (1879–1955) appeared on the 1968 Israeli 5 pound note. Einstein not only invented the theories of special relativity and general relativity, but also made fundamental contributions to the beginnings of quantum theory.

Science was constructed against a lot of nonsense.

> — *Albert Libchaber*

Where chaos begins, classical science stops.

> — *James Gleick*
> In *Chaos: Making a New Science*, 1987

The aim of science is not to open the door to everlasting wisdom, but to set to a limit on everlasting error.

> — *Bertolt Brecht*
> In *Life of Galileo*, 1938–1939

The progress of Science consists in observing interconnections and in showing with a patient ingenuity that the events of this ever-shifting world are but examples of a few general relations, called laws. To see what is general in what is particular, and what is permanent in what is transitory, is the aim of scientific thought.

> — *Alfred North Whitehead*
> In *An Introduction to Mathematics*

The role of science, like that of art, is to blend proximate imagery with more distant meaning, the parts we already understand with those given as new into larger patterns that are coherent enough to be acceptable as truth. Biologists know this relation by intuition during the course of fieldwork, as they struggle to make order out of the infinitely varying patterns of nature.

> — *Edward O. Wilson*
> In *In Search of Nature*, 1996

The materials of science are the materials of life itself. Science is part of the reality of living; it is the what, the how, and the why of everything in our experience.

> — *Rachel Carson, 1952*
> Quoted in *The House of Life* by Paul Brooks

Science is vastly more stimulating to the imagination than are the classics.

> — *John Haldane*
> In *Daedalus*, 1924

Science… commits suicide when it adopts a creed.

> — *Thomas Huxley*
> In *Darwiniana*, "The Darwin Memorial," 1893

He who made us would have been a pitiful bungler, if he had made the rules of moral conduct a matter of science.

> — *Thomas Jefferson*
> In a letter to Peter Carr, 1787

Putting on the spectacles of science in expectation of finding the answer to everything looked at signifies inner blindness.
> —*J. Frank Dobie*
> In *The Voice of the Coyote*, 1949

Scientists are Peeping Toms at the keyhole of eternity.
> —*Arthur Koestler*
> Quoted in *Archaeology* by David Thomas
> *The truly great scientist enters the keyhole.*

Physics

At any rate, I am convinced that He [God] does not play dice.
> —*Albert Einstein*
> In a letter to Max Born,
> December 4, 1926
> *Stating his displeasure with
> the idea of quantum mechanics.
> Also quoted as "God does not play dice
> with the world."*

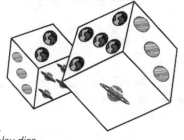

Who are you to tell God what to do?
> —*Attributed to Niels Bohr*
> *In reply to Einstein. Also quoted as "Albert, don't tell God what to do."*

God plays dice with the universe, but they're loaded dice. And the main objective of physics now is to find out by what rules were they loaded and how can we use them for our own ends.
> —*Joseph Ford*
> Quoted in *Chaos: Making a New Science* by James Gleick

God not only plays dice, He also sometimes throws the dice where they cannot be seen.
> —*Stephen Hawking*
> In *Nature*, 1975

Disorder increases with time because we measure time in the direction in which disorder increases.
> — *Stephen Hawking*
> *The simplest and most accurate explanation for "why entropy."*

What makes planets go around the sun? At the time of Kepler some people answered this problem by saying that there were angels behind them beating their wings and pushing the planets around an orbit. As you will see, the answer is not very far from the truth. The only difference is that the angels sit in a different direction and their wings push inward.
> — *Richard Feynman*
> In *Character Of Physical Law*, 1967
> *A planet orbits because it accelerates in the direction of the sun's gravity. As far as we know, gravity could be caused by tiny angels beating their wings.*

Every particle of matter is attracted by or gravitates to every other particle of matter with a force inversely proportional to the squares of their distances.
> — *Sir Isaac Newton*
> In *Principia*, 1687
> *This is Newton's famous expression for the universal law of gravitation.*

Every body continues in its state of rest or uniform motion in a straight line, except in so far as it doesn't.
> — *Sir Arthur Stanley Eddington*
> In *The Nature of the Physical Word*, 1933

We already know the physical laws that govern everything we experience in everyday life... It is a tribute to how far we have come in theoretical physics that it now takes enormous machines and a great deal of money to perform an experiment whose results we cannot predict.
> — *Stephen Hawking*
> In a 1980 lecture titled "Is the End in Sight for Theoretical Physics?"
> *The difficulty arises in trying to apply the basic physical laws to a realistic system. That is the place of the engineer and scientist, and there shall be no end to the need for engineers and scientists.*

In a few years, all great physical constants will have been approximately estimated, and that the only occupation which will be left to men of science will be to carry these measurements to another place of decimals.

— *James Clerk Maxwell*
In *Scientific Papers*, October 1871
Many new physical constants, such as Planck's constant, have been discovered since Maxwell's comment.

Oh, that stuff! We never bother with that in our work.

— *Ernest Rutherford*
In conversation, quoted in *The Boy I Left Behind Me* by Stephen Leacock.
In reply to what he thought of Albert Einstein's theory of relativity.

X-rays will prove to be a hoax.

— *William Thomson Kelvin*

It was Einstein who made the real trouble. He announced in 1905 that there was no such thing as absolute rest. After that there never was.

— *Stephen Leacock*
In *The Boy I Left Behind Me*, 1947
Corollary: Heisenberg announced that while you may have absolute rest, you will have no idea as to your location.

Physics becomes in those years the greatest collective work of science—no, more than that, the great collective work of art of the twentieth century.

— *Jacob Bronowski*
In *The Ascent of Man*, 1973
Referring to the period around the turn of the century marked by the discovery of atomic structure and the development of quantum theory.

Cold isn't a thing at all; it's merely the absence of heat, as dark is the absence of light. In a place where there is absolutely no heat it must be absolutely cold, and it couldn't get any colder.

— *Leslie Greener*
In *Moon Ahead*, 1951
An apt description of the concept of absolute temperature where nothing can be colder than −273°C.

It is wrong to think that the task of physics is to find out how Nature is. Physics concerns what we say about Nature.
> — *Niels Bohr*

Physics is much too hard for physicists.
> — *David Hilbert*
> Quoted in *Hilbert* by C. Reid
> *The implication is that physics should be reserved for mathematicians.*

Physicists can only think the same damn thing over and over.
> — *Richard Feynman*
> In *QED: The Strange Theory of Light and Matter*, 1990
> *Remarking about the similarity between the standard model of physics and the models for QED (quantum electrical dynamics).*

Berthelot had predicted that in a hundred years of physical and chemical science man would learn to know the atom, and that with this knowledge he would be able, at his will, to dim, extinguish or relight the sun like a Carcel lamp.
> — *Edmond de Goncourt and Jules de Goncourt*
> In "Journals," April 7, 1869
> *The prediction is surprisingly accurate. Although we cannot dim the sun, we can create and control our own little suns within reactors on Earth.*

The interaction between a husband and wife is significantly complicated by having the wife's lover hiding in the bedroom. This is called the three-body problem.
> — *H. Pierre Noyes*
> In a physics lecture at Stanford University, 1995
> *The interaction between three masses is also known as the three-body problem and has been one of the classic "unsolvable" problems since Newton's time.*

Put your hand on a hot stove for a minute, and it seems like an hour. Sit with a pretty girl for an hour, and it seems like a minute. *That's* relativity.
> — *Albert Einstein*
> In the *News Chronicle*, March 14, 1949
> *Einstein won the Nobel Prize for the theory of special relativity.*

If you see a formula in the *Physical Review* that extends over a quarter of a page, forget it. It's wrong. Nature isn't that complicated.
> — *Bernd T. Matthias*
>> *And if you see an extensive equation in an engineering journal, then forget it because you will never be able to apply it for practical purposes.*

Since the mathematicians have invaded the theory of relativity, I do not understand it myself anymore.
> — *Albert Einstein*
>> In "To Albert Einstein's Seventieth Birthday" by A. Sommerfelt in *Albert Einstein, Philosopher-Scientist* by Paul Schilpp

This perpetual motion machine that Lisa made is a joke, it just keeps going faster and faster…Lisa, in this house we obey the laws of thermodynamics!
> — *Matt Groening*
>> In *The Simpsons*, spoken by Homer Simpson

The infinite! No other question has ever moved so profoundly the spirit of man.
> — *David Hilbert*

[Overheard in a] Theoretical Physics Lab: "Maybe I can't fully grasp infinity, but I can comprehend more of it than you can!"
> — *Bob Thaves*
>> As a cartoon caption in *Frank & Ernest*, 1998

The speed of light is 3×10^8 m/s. It's not just a good idea, it's the law!
> — *Anonymous*
>> *This is the maximum speed that light can travel in any medium.*

There was a young lady named Bright,
Whose speed was far faster than light;
   She set out one day
   In a relative way,
And returned home the previous night.
> — *Arthur Henry Reginald Buller*
>> In *Punch*, December 19, 1924

When Newton saw an apple fall, he found…
A mode of proving that the earth turned round
In a most natural whirl, called gravitation;
And thus is the sole mortal who could grapple
Since Adam, with a fall or with an apple.
> — *George Gordon, Lord Byron*

I know that this defies the law of gravity, but, you see, I never studied law.
> — *Tex Avery*
> In the cartoon **Bugs Bunny**, spoken by Bugs Bunny

Q: What is the definition of a tachyon?
A: It's a gluon that's not completely dry.
> — *Anonymous*
> *A tachyon is a mythical particle that travels faster than light and, thus, has imaginary mass. A gluon is conjectured to be the force that binds quarks together to form protons.*

Q: Mother Nature's floor is *filthy*. Why doesn't she clean up this mess?
A: Nature *abhors* a vacuum.
> — *Mike Peters*
> As a cartoon caption in "Mother Goose & Grimm," 1998
> *And it is this filth that makes science (and life) interesting.*

BRICK: Well, they say nature hates a vacuum, Big Daddy.
BIG DADDY: That's what they say, but sometimes I think that a vacuum is a hell of a lot better than some of the stuff that nature replaces it with.
> — *Tennessee Williams*
> In *Cat on a Hot Tin Roof*, 1955

Weights and Measures

Measure what is measurable, and make measurable what is not so.
> — *Galileo Galilei*
> Quoted in "Mathematics and the Laws of Nature" by H. Weyl in *The Armchair Science Reader* by I. Gordon and S. Sorkin

Let us raise a standard to which the wise and honest can repair.
> — *George Washington*
> Inscribed over the entrance to the U.S. Commerce Building.

"This country is not tooled to go metric," West explains. What happens if American astronauts need to work on another country's modules? "We'll use Russian tools," he says.
> — *Phil Scott*
> In *Scientific American*, 1999
> *Phil West, referred to in the quote, is the NASA project manager of space-walk tools for the International Space Station.*

Heaviness is the intensity with which matter fills space.
> —*Julius Weisbach*
> In *Mechanics of Engineering, Theoretical Mechanics*, 1878

Do not give short measure and weight.
> — *The Koran 11:84*

My love is my weight.
> — *Saint Augustine*

The French kilogramme = 15,433.6 grains
> — *George Fownes*
> In *Elementary Chemistry, Theoretical and Practical*, 1855.
> *In 1855 the kilogram had not yet reached acceptance beyond France. For British chemists, the grain was the preferred measure of weight.*

Nearly all civilized nations have at some time employed a unit of length the name of which bore the same significance as does *foot* in English... But, as might have been expected, no two peoples have agreed in the length of their standard. Thus, the Macedonian foot was 14.08 inches and the Sicilian 8.75 inches... In Europe during the Middle Ages almost every town had its own characteristic foot.
> — *Robert Millikan and Henry Gale,*
> In *A First Course in Physics*, 1906
> *Newton talks about the circumference of the Earth as 123,249,600 Paris feet. The actual circumference is roughly 131,250,000 English feet.*

Materials Science

All things are Atoms: Earth and Water, Air and Fire.
> —*John Updike*
> In "Dance of the Solids," *Scientific American*, 1969

They [atoms] move in the void and catching each other up jostle
together, and some recoil in any direction that may chance, and others
become entangled with one another in various degrees according to
the symmetry of their shapes and sizes and positions and order, and
they remain together and thus the coming into being of composite
things is effected.
> — *Simplicius, 530 B.C.E*
> Quoted in Cyril Bailey's *The Greek Atomists and Epicurus*

Hence, Empresslike, *Ceramics* tend to be
Resistant, porous, brittle, and refractory.
> —*John Updike*
> In "The Dance of the Solids," *Scientific American*, 1969

Erwin Schroedinger
appears on the Austrian
1000 Schilling note.
Schroedinger was
one of the primary
developers of the
quantum theory.

When it comes to atoms, language can be used only as in poetry. The poet,
too, is not nearly so concerned with describing facts as with creating images.
> — *Niels Bohr*

The electron is not as simple as it looks.
> — *Lawrence Bragg*

We have learned that matter is weird stuff. It is weird enough, so that it does not limit God's freedom to make it do what he pleases.

— *Freeman J. Dyson*
In *Infinite in All Directions: Gifford lectures given at Aberdeen, Scotland, April–November 1985*, 1988

In Matter [is] the promise and potency of all forms of life.

— *George Bernard Shaw*
In the preface to *Back to Methuselah*, 1921

Surely Allah does not do injustice to the weight of an atom.

— *The Koran 4:40*

The house of tomorrow may have plastic walls and floors.

— *American Educator Encyclopedia, 1952*
If one includes acoustic ceiling tiles and vinyl floors, then this has already happened.

Electrostrictors all are ceramic;
with voltage aid, tiny or gigantic.
Despite high voltage and incredible strain
during large deflections, they feel no pain.
Yet when temperature is thus applied,
sensitive electrostrictors subside.
Cold electrostrictors need not alarm-
they're only shivering to keep warm.

— *M. Valere Masingill*
In "Ode to the Electrostrictor" in *Dynamic Actuation and Control with Electrostrictors by Michael Fripp*, 1995
Electrostrictors are a type of ceramic that changes size under an applied electric field and are useful for vibration control and for precision placement.

I don't like electrons; they've always had a negative influence on society.

— *Chris Lipe*

Chemistry

Organic chemistry is the chemistry of carbon compounds. Biochemistry is the study of carbon compounds that crawl.
> — *Mike Adams*

All substances are poisons: there is none which is not a poison. The right dose differentiates a poison from a remedy.
> — *Paracelsus, circa 1500*
> *Often paraphrased as "the dose makes the poison."*

The term *element* is applied in chemistry to those forms of matter which have hitherto resisted all attempts to decompose them. Nothing is ever meant to be affirmed concerning their real nature; they are simply elements to us at the present time; hereafter, by new methods of research, or by new combinations of those already possessed by science, many of the substances which now figure as elements may possibly be shown to be compounds; this has already happened, and may again take place.
> — *George Fownes*
> In *Elementary Chemistry, Theoretical and Practical*, 1855
> *The text continues by listing the 62 elements known at the time. Today there are more than 112 known elements. However, the elements of the periodic table are no longer elements in this sense of the definition because they can be subdivided into smaller particles.*

Oxygen: the great constituent of the ocean and the atmosphere.
> — *George Fownes*
> In *Elementary Chemistry, Theoretical and Practical*, 1855
> *The text orders elements by their importance instead of by the modern periodic table. The discussion, thus, leads with oxygen.*

We have absolutely no means at our disposal for deciding [the nature of elements] which remains at the present day in the same state as when it first engaged the attention of the Greek philosophers, or perhaps that of the sages of Egypt and Hindostan long before them.
> — *George Fownes*
> In *Elementary Chemistry, Theoretical and Practical*, 1855
> *We have subsequently been able to divide elements into subatomic particles and to achieve a good understanding of basic atomic behavior.*

Life exists in the universe only because the carbon atom possesses certain exceptional properties.

— *James Jeans*
In *The Mysterious Universe*

There is not a law under which any part of this universe is governed that does not come into play in the phenomena of the chemical history of a candle.

— *Michael Faraday*
In his children's lecture "The Chemical History of a Candle," 1860

There is hardly a boy or girl alive who is not keenly interested in finding out about things. And that's exactly what chemistry is: finding out about things—finding out what things are made of and what changes they undergo. What things? Any thing! Every thing!

— *Robert Brent*
In *The Golden Book of Chemistry Experiments*, 1960 and quoted by Ken Silverstein in *Harpers Magazine*

Chemistry means the difference between poverty and starvation and the abundant life.

— *Robert Brent*
In *The Golden Book of Chemistry Experiments*, 1960 and quoted by Ken Silverstein in *Harpers Magazine*

We boil at different degrees.

— *Ralph Waldo Emerson*
In *Society and Solitude*, 1870

There's nothing colder than chemistry.

— *Anita Loos*
In *Kiss Hollywood Good-by*, 1974

Every attempt to employ mathematical methods in the study of chemical questions must be considered profoundly irrational and contrary to the spirit of chemistry.... if mathematical analysis should ever hold a prominent place in chemistry—an aberration which is happily almost impossible— it would occasion a rapid and widespread degeneration of that science.

— *Auguste Comte*
In *Cours de Philosophie Positive*, 1830

It is disconcerting to reflect on the number of students we have flunked in chemistry for not knowing what we later found to be untrue.
> — *Robert L. Weber*
> In *Science With a Smile*, 1992

The chemical purity of the air is of no importance.
> — *L. Erskine Hill*
> Quoted in *The New York Times*, 1912

Shelley and Keats were the last English poets who were at all up to date in their chemical knowledge.
> —*John Haldane*
> In *Daedalus or Science and the Future*
> *This is a statement not so much of the knowledge of Shelley and Keats but rather of the explosive growth in chemical knowledge.*

Chemists are, on the whole, like physicists, only 'less so.' They don't make quite the same wonderful mistakes, and much of what they do is an art, related to cooking, instead of a true science. They have their moments, and their sources of legitimate pride. They don't split atoms, as the physicists do. They join them together, and a very praiseworthy activity that is.
> — *Anthony Standen*
> In *Science Is a Sacred Cow*, 1958

All theoretical chemistry is really physics; and all theoretical chemists know it.
> — *Richard Feynman*

It took the mob only a moment to remove his head; a century will not suffice to reproduce it.
> —*Joseph Louis Lagrange*
> *Speaking about Antoine Lavoisier, the founder of modern chemistry, who was guillotined during the French Revolution.*

Water is H_2O, hydrogen two parts, oxygen one, but there is also a third thing, that makes it water and nobody knows what that is.
> — *D.H. Lawrence*
> In *Pansies*, "The Third Thing"
> *The missing part is the bonding of the orbitals.*

He thought the formula for water was H-I-J-K-L-M-N-O (H-to-O).
— *Anonymous*

Remember, if you're not part of the solution, then you're part of the precipitate!
— *Eric Desch*

He doubted the existence of the Deity but accepted Carnot's cycle, and he had read Shakespeare and found him weak in chemistry.
— *H.G. Wells*
In "The Lord of the Dynamos" in *Short Stories*

Sir Humphry Davy
Abominated gravy.
He lived in the odium
Of having discovered Sodium.
— *Edmund Clerihew Bentley*
In the *Biography for Beginners*, 1905

Biology

Biology is the only science in which multiplication means the same thing as division.
— *Anonymous*

Know thyself.
— *Carolus Linnaeus*
In *Systema Naturae*, 1758
In his catalogue of all of the known species, Linnaeus also included a description of the animal. This is his description of Homo sapiens, a.k.a. humanity. Variations on this phrase have existed since antiquity.

Classification is a very human thing to do. By grouping, we can generalize and predict. Classification helps us to organize what we know, and it makes us better biologists.
— *Donald P. Abbott*
In *Observing Marine Invertebrates: Drawings from the Laboratory*, 1987

You'll be tempted to grouse about the instability of taxonomy; but stability occurs only where people stop thinking and stop working.
— *Donald P. Abbott*
In *Observing Marine Invertebrates: Drawings from the Laboratory*, 1987

The blood in the animal body is impelled in a circle and is in a state of ceaseless motion…and that is the sole and only end of the motion and contraction of the heart.
— *William Harvey*
In *On the Motion of the Heart and Blood*, 1628

The greatest single achievement of nature to date was surely the invention of the molecule of DNA.
— *Lewis Thomas*
In *The Medusa and the Snail*, 1979

We have discovered the secret of life!
— *Francis Crick*
In *The Double Helix, 1968*
Spoken after excitedly bursting into a Cambridge pub with James Watson to celebrate the fact that they had unraveled the structure of DNA.

Man is like every other species in being able to reproduce beyond the carrying capacity of any finite habitat. Man is like no other species in that he is capable of thinking about this fact and discovering its consequences.
— *William R. Catton, Jr.*
In *Overshoot: The Ecological Basis of Revolutionary Change*, 1980

Our biological nature does not stand in the way of social reform.
— *Steven J. Gould*
In "Biological Potentiality vs. Biological Determinism" in *Ever Since Darwin*, 1992

This is the essence of the matter as I understand it: culture is ultimately a biological product.
— *Edward O. Wilson*
In *In Search of Nature*, 1996

It would appear that the most frequent and fatal diseases of today are due to the "wear and tear" of modern life.

> — Scientific American
> Reporting on the work of Dr. Hans Selye, 1949
>
> *Selye discovered the existence of stress-related disease due to his poor rat handling techniques during an attempt to study the effect of estrogen on rats. Selye stressed the rats of both the control and experimental groups with his poor rat handling technique. He "would try to inject the rats, miss them, drop them, spend half the morning chasing the rats around the room or vice versa, flailing with a broom to get out from behind the sink, and so on." (from* Why Zebras Don't Get Ulcers *by Robert Sapolsky)*

The real revolution in medicine…began with the destruction of dogma. It was discovered, sometime in the 1830s, that the greater part of medicine was nonsense.

> — *Lewis Thomas*
> In *The Medusa in the Snail*, 1979

If the brain were so simple we could understand it, we would be so simple we couldn't.

> — *Lyall Watson*

I have had several gentlewomen in my house, who were keen on seeing the little eels in vinegar, but some of them were so disgusted at the spectacle, that they vowed they'd never use vinegar again. But what if one should tell such people in the future that there are more animals living in the scum on the teeth in a man's mouth, than there are men in a whole kingdom?

> — *Anton van Leeuwenhoek*
> *Dutch merchant and discoverer of the microbial world through microscopes.*

...and when I'm gone, I would like to donate my body to a biology class.

Physiological experiment on animals is justifiable for real investigation, but not for mere damnable and detestable curiosity.

> — *Charles Robert Darwin*
> In a letter to E. Ray Lankester, 1900

Every species lives a life unique to itself, and every species dies in a different way.
> — *Edward O. Wilson*
> In *The Diversity of Life*, 1992

There's no substitute for fine forceps. None.
> — *Donald P. Abbott*
> In *Observing Marine Invertebrates: Drawings from the Laboratory*, 1987
> *Speaking of animal dissections. Even in biology, the proper tools are essential.*

Cell and tissue, shell and bone, leaf and flower, are so many portions of matter, and it is in obedience to the laws of physics that their particles have been moved, molded and conformed... Problems of form are in the first instance mathematical problems, their problems of growth are essentially physical problems, and the morphologist is, *ipso facto*, a student of physical science.
> — *D'Arcy Wentworth Thompson*
> In *On Growth and Form*, 1917

The safest and the most prudent of bets to lay money on is surprise. There is a very high probability that whatever astonishes us in biology today will turn out to be usable, and useful, tomorrow.
> — *Lewis Thomas*
> In *The Medusa and the Snail*, 1979
> *This perspective applies to most areas of science.*

We gave some account a few weeks ago of the astonishing case of Mr. Gage, foreman of the railroad in Cavendish, who in preparing a charge for blasting a rock had an iron bar driven through his head, entering through his cheek and passing out at the top with a force that carried the bar several yards, after performing its wonderful journey through skull and brains. We refer to this case again to say that the patient not only survives but is much improved. He is likely to have no visible injury but the loss of an eye.
> — *In* The Mercury *of Woodstock Vt., 1848*
> *Phineas Gage survived for 12 years but with a radically altered personality. His case is still studied today as a model of cerebral function.*

[Louis Pasteur's]... theory of germs is a ridiculous fiction. How do you think that these germs in the air can be numerous enough to develop into all these organic infusions? If that were true, they would be numerous enough to form a thick fog, as dense as iron.

— *Pierre Pochet*
In *The Universe: The Infinitely Great and the Infinitely Small*, 1872

What remains to be said is of so novel and unheard of a character that I not only fear injury to myself from the envy of a few, but I tremble lest I have mankind at large for my enemies, so much to wont and custom that become as another nature, and doctrine once sown that hath struck deep root, and respect for antiquity, influence all men.

— *William Harvey*
In *On the Motion of the Heart and Blood*, 1628
Introduction to the book that was described by Sir John Simon as the most important ever made in physiological science and revolutionized the way that science considered the blood and heart.

A zygote is a gamete's way of producing more gametes. This may be the purpose of the universe.

— *Robert Heinlein*
In *Time Enough for Love*, 1973

A band of bacterial brothers
Swigging ATP with some others,
In a jocular fit,
They laughed 'til they split
Now they're all microbial mothers.

— *Richard Cowen*
In *History of Life*, 1989
ATP (adenosine triphosphate) is the molecule that cells use to store energy. This quote refers to the process by which microbes, another word for single-celled organisms such as bacteria, reproduce by dividing one cell into two independent cells.

Glycocalyx, the slime on bacteria
Helps them live in a bovine interior
Powerful drugs
Can't get at these bugs
It's clearly a substance superior.

— *Richard Cowen*
In *History of Life*, 1989

Animal Behavior

All thoughts of a turtle are turtles, and of a rabbit, rabbits.

> — *Ralph Waldo Emerson*
> In *The Natural History of Intellect*, 1893
> *Some thoughts of rabbits are of wolves.*

Be a good animal, true to your instincts.

> — *D.H. Lawrence*

The choice of sampling method restrict[s] the kinds of behavior processes that can be studied.

> — *Jeanne Altmann*
> In "Observational study of behavior: sampling methods," in *Animal Behavior*, 1974
> *What you record and how you record it will have a profound impact on the conclusions that can be drawn.*

An unambiguous formulation of a behavioral question is critical to the choice of an appropriate sampling technique.

> — *Jeanne Altmann*
> In "Observational sampling methods for insect behavioral ecology," in *The Florida Entomologist,* 1981

Some men have strange ambitions. I have one:
To make a naturalist without a gun.

> — *E. Selous*
> In *The Bird Watcher in the Shetlands with Some Notes on Seals— and Digressions*, 1905
> *Early naturalists were more concerned with collecting specimens than with observing the animal's behavior.*

At last such field studies have been put on a sound basis which should result in the hunting of information rather than specimens.

> — *H. C. Bingham*
> In *Gorillas in a Native Habitat,* 1932

Get the experience of looking at fresh things. If you watch live animals, you gain clearer insights in shorter time than you would watching dead animals for much longer.

> — *Donald P. Abbott*
> In *Observing Marine Invertebrates: Drawings from the Laboratory*, 1987

Individual recognition of animals is essential in a detailed study of social behavior.

> — *George B. Schaller*
> In "Field Procedures" in *Primate Behavior: Field Studies of Monkeys and Apes* by Irvine Devore, 1965

Behavioral biologists rarely recognized behavioral contributions to conservation other than captive breeding and reintroduction programs of endangered species.

> — *J. R. Clemmons and R. Buchholz*
> In *Behavioral Approaches to Conservation in the Wild*, 1997
>
> *It is only recently that behavioral biologists have realized that they have more to contribute to conservation. Knowledge of the animals' natural behavior is very important to the ability to preserve species.*

The biologist, educated to respect "hard data," ...usually views behavioral work as occupying the "soft" fringe of biology (meaning that area which abuts on psychology and is therefore barely respectable).

> — *D. E. Gaskin*
> In *The Ecology of Whales and Dolphins,* 1982

Flexibility may well be the most important determinant of human consciousness.

> — *Steven J. Gould*
> In "Biological Potentiality vs. Biological Determinism"
> in *Ever Since Darwin*, 1992

Man cannot live by milk alone.

> — *Harry Harlow*
> In "Affectional Responses in the Infant Monkey" in *Science*, volume 130, August 21, 1959
>
> *Harry Harlow is biologist who performed experiments showing that young monkeys would rather choose a soft, terry-cloth "mother" who did not give milk over a wire cage "mother" who would give milk.*

As Benjamin Franklin said:"We must all hang together, or assuredly
we shall all hang separately." Functioning societies may require
reciprocal altruism.

> — *Steven J. Gould*
> In "Biological Potentiality vs. Biological Determinism"
> in *Ever Since Darwin*, 1992
> *Reciprocal altruism occurs when animals perform helpful acts towards
> others because the recipient will repay the acts in the future.*

I would give my life for two brothers or eight cousins.

> — *W.D. Hamilton, 1964*
> *Explanation of his theory of altruism based upon kin selection noting that
> you share equal genetics with two brothers or eight cousins.*

The statement that humans are animals does not imply that our
specific patterns of behavior and social arrangements are in any way
directly determined by our genes. *Potentiality* and *determination* are
different concepts.

> — *Steven J. Gould*
> In "Biological Potentiality vs. Biological Determinism"
> in *Ever Since Darwin*, 1992

Behaviorism is the art of pulling habits out of rats.

> — *Anonymous*

Evolution

The ape, vilest of beasts, how like to us!

> — *Quintus Ennius*
> Quoted by Marcus Tillius Cicero in *De Senectute, IV*, circa 70 B.C.E.

The hallmark of life is this: a struggle among an immense variety of
organisms weighing next to nothing for a vanishingly small amount
of energy.

> — *Edward O. Wilson*
> In *The Diversity of Life*, 1992

I have called this principle, by which each slight variation, if useful, is preserved, by the term Natural Selection.

> — *Charles Robert Darwin*
> In *The Origin of Species*, 1859

Early ethologists often assumed that natural selection would produce animals that sacrificed personal reproductive success for the general benefit of their species.

> — *J. Alcock*
> In *Animal Behavior: an Evolutionary Approach,* 1993
>
> *This concept is called "group selection." It is now understood that natural selection works through genetics and therefore acts on individuals, not groups. Individuals do not sacrifice personal reproductive success for the benefit of the species, only of their own kin.*

We have unmistakable proof that throughout all past time, there has been a ceaseless devouring of the weak by the strong.

> — *Herbert Spencer*
> In *First Principles*, 1861

Here, you see, it takes all the running you can do, to keep in the same place.

> — *Lewis Carroll*
> In *Alice Through the Looking Glass*, 1872
>
> *This is the evolutionary principle known as the Red Queen Hypothesis. It is the Red Queen who is speaking to Alice. The hypothesis states that because the environment is constantly changing, organisms have to be constantly changing in order to remain as fit for the environment as they had been previously.*

The Simiadae then branched off into two great stems, the New World and Old World monkeys; and from the latter at a remote period, Man, the wonder and the glory of the universe, proceeded.

> — *Charles Robert Darwin*
> In *The Descent of Man*, 1871

Is man an ape or an angel? I, my lord, I am on the side of the angels. I repudiate with indignation and abhorrence those new fangled theories.

> — *Benjamin Disraeli, Earl of Beaconsfield*
> In a speech at Oxford Diocesan Conference, November 25, 1864
>
> *In reply to Charles Darwin's theories of evolution.*

I believe that our Heavenly Father invented man because he was disappointed in the monkey.
> — *Mark Twain*
>> *Or was it the other way around?*

Evolution is chaos with feedback.
> — *Joseph Ford*
>> Quoted in *Chaos* by James Gleick

Natural selection is a mechanism for generating an exceedingly high degree of improbability.
> — *Ronald Fisher*

The capacity to blunder slightly is the real marvel of DNA. Without this special attribute, we would still be anaerobic bacteria and there would be no music.
> — *Lewis Thomas*
>> In *The Medusa and the Snail*, 1979

We are here by the purest chance, and by mistake at that.
> — *Lewis Thomas*
>> In *The Medusa and the Snail*, 1979
>> *To err is human but erring is biological as well. Without replication errors, there would be no evolution and we would all be bacteria.*

Biology needs a better word than "error" for the driving force in evolution. Or maybe "error" will do after all, when you remember that it came from an old root meaning to wander about, looking for something.
> — *Lewis Thomas*
>> In *The Medusa and the Snail*, 1979

Change is the essential process of all existence.
> — *Oliver Crawford*
>> In *Star Trek: Let That Be Your Last Battlefield*, 1969

By natural selection our mind has adapted itself to the conditions of the external world. It has adopted the geometry most advantageous to the species or, in other words, the most convenient.
> — *Jules Henri Poincaré*
>> In *Science and Method*, 1908

Evolution on a large scale unfolds, like much of
human history, as a succession of dynasties.
— *Edward O. Wilson*
In *The Diversity of Life*, 1992

It was a misfortune for the living world in
particular, many scientists believe, that a
carnivorous primate and not some more benign
form of animal made the breakthrough.
— *Edward O. Wilson*
In *In Search of Nature*, 1996
*Speaking of man's breakthrough to
intelligence and, hence, to global dominance.*

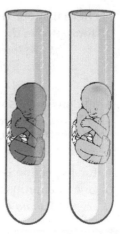

The advantage of
being a test-tube baby:
A Womb with a View

Darwin's dice have rolled badly for Earth.
— *Edward O. Wilson*
In *In Search of Nature*, 1996
*Referring to the fact that the one intelligent species
on Earth is a carnivorous primate.*

Every species of plant and animal is determined by a pool of germ plasm
that has been most carefully selected over a period of hundreds of
millions of years. We can understand now why it is that mutations in
these carefully selected organisms almost invariably are detrimental. The
situation can be suggested by a statement by Dr. J.B.S. Haldane: "My clock
is not keeping perfect time. It is conceivable that it will run better if I
shoot a bullet through it; but it is much more probable that it will stop
altogether." Professor George Beadle, in this connection, has asked:
"What is the chance that a typographical error would improve Hamlet?"
— *Linus Pauling*
In *No more War!*, 1958

Pouter, tumbler and fantail are from the same source;
The racer and hack may be traced to one horse;
So men were developed from monkeys, of course,
    Which nobody can deny.
— *Lord Charles Neaves*
In "The Origin of Species"

[A] curious aspect of the theory of evolution is that everybody thinks
he understands it.
>   —*Jacques Monod*
>   In *On the Molecular Theory of Evolution*, 1974

Amoebas at the start
   Were not complex;
They tore themselves apart
   And started Sex.
>   —*Arthur Guiterman*
>   In "Sex"
>   *This is actually asexual reproduction. Sexual reproduction requires cells to be
>   "put together" rather than "torn apart."*

One day the zoo-keeper noticed that the orangutan was reading
two books—the Bible and Darwin's Origin of Species. In surprise,
he asked the ape, "Why are you reading both those books?"
   "Well," said the orangutan, "I just wanted to know if I was my
brother's keeper, or my keeper's brother."
>   —*Anonymous*
>   *Moral: We could ask the same question.*

Cloning

The cloning of humans is on most of the lists of things to worry about
from Science, along with behaviour control, genetic engineering,
transplanted heads, computer poetry and the unrestrained growth of
plastic flowers.
>   —*Lewis Thomas*
>   In *The Medusa and the Snail*, 1979

Cloning is the most dismaying of prospects, mandating as it does the
elimination of sex with only a metaphoric elimination of death as
compensation.
>   —*Lewis Thomas*
>   In *The Medusa and the Snail*, 1979

What nature does blindly, slowly and ruthlessly, man may do providently, quickly and kindly.
> — *Francis Galton*

For the first time in all time, a living creature understands its origin and can undertake to design its future.
> — *Robert Sinsheimer, 1969*
> *Remarking that biologists can now repair genes.*

There is no way to tell what the effects of eliminating one gene from the human population would be.
> — *Hamilton Smith*
> *Cautioning against manipulating DNA.*

To do the thing [cloning] properly, with any hope of ending with a genuine duplicate of a single person, you really have no choice. You must clone the world, no less.
> — *Lewis Thomas*
> In *The Medusa and the Snail*, 1979
> *Cloning only reproduces the genetics. A person's developmental environment also plays a significant role in that person's personality.*

I sleep better at night knowing that scientists can clone sheep.
> —*Jeff Ayers, 1999*

Claude Bernard...announced that with a hundred years more of physiological knowledge we would be able to make the organic law ourselves—to manufacture human life, in competition with the Creator.

We believe that at that particular stage of scientific development, the good Lord, with a flowing white beard, will arrive on Earth with his chain of keys and will say to humanity, just as they do at the Art Gallery at five o'clock, "Gentlemen, it's closing time."
> — *Edmond de Goncourt and Jules de Goncourt*
> In *Journals*, April 7, 1869

The cloning of mammals... is biologically impossible.
> —*James McGrath and Davor Solter*
> In *Science*, December 1984

Archaeology

You have to know the past to understand the present.
> — *Carl Sagan*
> In *Cosmos*, 1993

To find old sites, you must look in old dirt.
> — *Jonathan O. Davis*

You may forget but let me tell you this: someone in some future time
will think of us.
> — *Sappho*

The destruction of a prehistoric site is permanent. Like Humpty Dumpty,
it cannot be reassembled.
> — *Louis Brennan*

Archaeologists derive scant comfort from the fact that over and above
the certainties of death and taxes, they are blessed with the additional
constant of a seemingly limitless quantity of shards to classify.
> — *Prudence M. Rice*
> Shards are broken pieces of pottery.

The probability of making surface finds decreases in inverse ratio to
the square of the distance between the ground and the end of the
searcher's nose.
> — *Louis Brennan*

An archaeologist is the best husband any woman can have: the older
she gets, the more interested he is in her.
> — *Agatha Christie*
> Quoted in news reports on March 9, 1954
> Christie later denied this remark. It is attributed to her by her second husband,
> Sir Max Mallowan.

Caves are wonderful places for lairs
For sabertooth tigers and bears
"But 'try and eject us!"
Said Homo erectus,
"We need this place for our heirs!"
> — *Richard Cowen*
> In *History of Life*, 1989

History records the names of royal bastards, but cannot tell us the origin of wheat.
> — *Jean Henri Fabre*

Biographical history... is still largely a history of boneheads; ridiculous kings and queens, paranoid political leaders, compulsive voyages, ignorant generals—the flotsam and jetsam of historical currents. The men [and women] who radically altered history, the great scientists and mathematicians, are seldom mentioned, if at all.
> — *Martin Gardner*

Anthropology

Anthropologists are a connecting link between poets and scientists; though their field-work among primitive peoples has often made them forget the language of science.
> — *Robert Graves*
> In a speech at the London School of Economics, December 6, 1963

Anthropologists are highly individual and specialized people. Each of them is marked by the kind of work he or she prefers and has done, which in time becomes an aspect of that individual's personality.
> — *Margaret Mead*

How can a modern anthropologist embark upon a generalization with any hope of arriving at a satisfactory conclusion? By thinking of the organizational ideas that are present in any society as a mathematical pattern.
> — *Edmund Leach*
> In *Rethinking Anthropology*, 1961

Anthropology is the science which tells us that people are the same the whole world over—except when they are different.
— *Nancy Banks-Smith*
Quoted in the *Guardian*, a London newspaper, July 21, 1988

Anthropology has always struggled with an intense, fascinated repulsion towards its subject… [The anthropologist] submits himself to the exotic to confirm his own inner alienation as an urban intellectual.
— *Susan Sontag*
Quoted in "Structure and Infrastructure in Primitive Society" by Neville Dyson-Hudson in *The Structuralist Controversy* by R. Macksey and E. Donato.

Research

Research is what I'm doing when I don't know what I'm doing.
— *Wernher von Braun*

If we knew what it was we were doing, it would not be called research, would it?
— *Albert Einstein*

Research is formalized curiosity. It is poking and prying with a purpose.
— *Zora Neale Hurston*
In *Dust Tracks on a Road*, 1942

Discovery consists of seeing what everybody has seen and thinking what nobody has thought.
— *Albert Szent-Gyorgyi*
Quoted in *The Scientist Speculates* by L.J. Good

Science is facts; just as houses are made of stone, so is science made of facts; but a pile of stones is not a house and a collection of facts is not necessarily science.
— *Henri Poincaré*
In *Science and Hypothesis*, 1913

Ask, and it shall be given you; seek, and ye shall find; knock, and it shall
be opened unto you.
— *Matthew 7:7*

When faced with a problem you do not understand, do any part of it you
do understand and then look at it again.
— *Robert Heinlein*

The by-product is sometimes more valuable than the product.
— *Havelock Ellis*
In *Little Essays on Love and Virtue*

All human knowledge thus begins with intuitions, proceeds thence to
concepts, and ends with ideas.
— *Immanual Kant*
Quoted in *Foundations of Geometry* by David Hilbert

It's not what you find, it's what you find out.
— *David Hurst Thomas*
In *Archaeology*, 1989
The data are not as important as the insightful interpretation of the data.

The most exciting phrase to hear in science, the one that heralds new
discoveries, is not "Eureka!" (I found it!) but "That's funny ..."
— *Isaac Asimov*

Each problem that I solved became a rule which served afterwards to
solve other problems.
— *René Descartes*
In *Discours de la Methode*, 1637

The outcome of any serious research can only be to make two questions
grow where only one grew before.
— *Thorstein Veblen*
In *The Place of Science in Modern Civilization and Other Essays*, 1919

Thought is only a flash between two long nights,
but this flash is everything.
— *Jules Henri Poincaré*

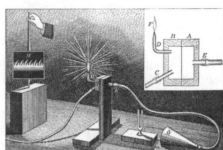

You know my methods, Watson.
— *Sir Arthur Conan Doyle*
In *The Memoirs of*
Sherlock Holmes, 1894

The absence of evidence is not
evidence of absence.
— *Carl Sagan*
Although spoken about the
search for extraterrestrial intelligence,
this philosophy is applicable to many fields of research.

Analysis of sounds with manometric flames from
First Course in Physics by Millikan and Gale, 1906,
pg 374

Mysteries are not necessarily miracles.
— *Johann Goethe*

Research is a way of taking calculated risks to bring about
incalculable consequences.
— *Celia Green*
In *The Decline and Fall of Science*, 1976

The way to do research is to attack the facts at the point of
greatest astonishment.
— *Celia Green*
In *The Decline and Fall of Science*, 1976

I love fools' experiments; I am always making them.
— *Charles Darwin*

The greatest reward lies in making the discovery; recognition can add
little or nothing to that.
— *Franz Ernst Neumann*

A wrench is a great tool, but don't try to drive a nail with it.
— *Anonymous*
A comment on those who solely rely upon a single approach.

The main cause of this unparalleled progress in physiology, pathology, medicine, and surgery has been the fruitful application of the experimental method of research, just the same method which has been the great lever of all scientific advance in modern times.
— *Dr. William H. Welch*
In *Argument Against Antivivisection Bill*, 1900

If you do something once, people will call it an accident. If you do it twice, they call it a coincidence. But do it a third time and you've just proven a natural law.
— *Grace Murray Hooper*
Quoted in *Mothers of Invention* by Ethlie Vare and Greg Ptacek.

May every young scientist remember and not fail to keep his eyes open for the possibility that an irritating failure of his apparatus to give consistent results may once or twice in a lifetime conceal an important discovery.
— *Patrick Blackett*

An enemy whose intentions one knows; is already half beaten.
— *Ou-Tse, 300 B.C.E.*
The most difficult part of research is discerning the interesting fundamental questions. Once the questions can be expressed, then the solution tends to be readily available.

The joy of suddenly learning a former secret and the joy of suddenly discovering a hitherto unknown truth are the same to me—both have the flash of enlightenment, the almost incredibly enhanced vision, and the ecstasy and euphoria of released tension.
— *Paul Halmos*
In *I Want to be a Mathematician*, 1985

The simplest solution is the best solution.
— *Craig Fischenich*

Good science is almost always so very simple. *After* it has been done by someone else, of course.

> — *L.L. Larison Cudmore*
> In *The Center of Life*, 1977

[Research is analogous] to seeing the peak of a mountain and trying to climb to the top. One establishes a base camp and begins scaling the mountain's sheer face, encountering obstacles at every turn, often retracing one's steps and struggling every foot of the journey. Finally when the top is reached, one stands examining the peak, taking in the view of the surrounding countryside and then noting the automobile road up the other side!

> — *Robert Kleinhenz*

Hell is truth seen too late.

> — *Jewish proverb*

Research! A mere excuse for idleness; it has never achieved, and will never achieve any results of the slightest value.

> — *Benjamin Jowett*

Aristotle maintained that women have fewer teeth than men; although he was twice married, it never occurred to him to verify this statement by examining his wives' mouths.

> — *Bertrand Russell*
> In Impact of Science on Society, 1952

Who gathers knowledge gathers pain.

> — *Ecclesiastes 1:18*

Behavior of Scientists

A scientist in his laboratory is not only a technician: he is also a child placed before natural phenomena which impress him like a fairy tale.

— *Marie Curie, 1893*
Quoted in *Madame Curie* by Eve Curie

Whatever a scientist is doing—reading, cooking, talking, playing— science thoughts are always there at the edge of the mind. They are the way the world is taken in; all that is seen is filtered through an everpresent scientific musing.

— *Vivian Gornick*
In *Women in Science*, 1983

A scientist or a writer is one who ruminates continuously on the nature of physical or imaginative life, experiences repeated relief and excitement when the insight comes, and is endlessly attracted to working out the idea.

— *Vivian Gornick*
In *Women in Science*, 1983

If I could remember the names of all these particles I'd be a botanist.

— *Enrico Fermi*
Quoted in *More Random Walks in Science* by R.L. Weber
Famous nuclear physicist referring to the myriad of subatomic particles.

But when the rigorous logic of the matter is not plain! Well, what of that? Shall I refuse my dinner because I do not fully understand the process of digestion?

— *Oliver Heaviside*
In *Electromagnetic Theory, Vol. II*, 1899
When criticized for using formal mathematical relations without understanding how they worked.

My goal is simple. It is complete understanding of the universe, why it is as it is and why it exists at all.

— *Stephen Hawking*

It was absolutely marvellous working for Pauli. You could ask him anything. There was no worry that he would think a particular question was stupid, since he thought *all* questions were stupid.

> — *Victor Weisskopf*
> In *American Journal of Physics*, 1977
> *Wolfgang Pauli was a Nobel Prize winning physicist who studied quantum mechanics.*

Someone remarked to me once: "Physicians shouldn't say, I have cured this man, but, this man didn't die under my cure." In physics too, instead of saying, I have explained such and such a phenomenon, one might say, "I have determined causes for it."

> — *George Christoph Lichtenberg*

Those who dwell as scientists…among the beauties and mysteries of the earth are never alone or weary of life. Those who contemplate the beauty of the earth find reserves of strength that will endure as long as life lasts.

> — *Rachel Carson*
> In *The Sense of Wonder*, 1956

One never notices what has been done; one can only see what remains to be done.

> — *Marie Curie*
> In a letter to her brother, March 18, 1894

Almost anyone can do science; almost no one can do good science.

> — *L.L. Larison Cudmore*
> In *The Center of Life*, 1977

One could not be a successful scientist without realizing that, in contrast to the popular conception supported by newspapers and mothers of scientists, a goodly number of scientists are not only narrow-minded and dull, but also just stupid.

> —*James D. Watson*
> In *The Double Helix*, 1968

In science it often happens that scientists say, "You know that's a really good argument; my position is mistaken," and then they would actually change their minds and you never hear that old view from them again. They really do it. It doesn't happen as often as it should, because scientists are human and change is sometimes painful. But it happens every day. I cannot recall the last time something like that happened in politics or religion.

> — *Carl Sagan*
> In a lecture, 1987

On 13 September 1765 people in fields near Luce, in France, saw a stone-mass drop from the sky after a violent thunderclap. The great physicist Lavoisier, who knew better than any peasant that this was impossible, reported to the Academy of Science that the witnesses were mistaken or lying. The Academy would not accept the reality of meteorites until 1803.

> — *Anonymous*

Not only was he [Newton] the greatest genius that ever existed, but also the most fortunate for we cannot find more than once a system of the world to establish.

> — *Joseph Louis Lagrange*
> Quoted in *The World's Greatest Books* by Arthur Mee and J.A. Hammerton
> *The same could be said about Einstein.*

If we evolved a race of Isaac Newtons, that would not be progress. For the price that Newton had to pay for being a supreme intellect was that he was incapable of friendship, love, fatherhood, and many other desirable things. As a man he was a failure; as a monster he was superb.

> — *Aldous Huxley*
> In an interview with J.W.N. Sullivan in *Contemporary Mind*, 1934

The modern physicist is a quantum theorist on Monday, Wednesday, and Friday and a student of gravitational relativity theory on Tuesday, Thursday, and Saturday. On Sunday he is neither, but is praying to his God that someone, preferably himself, will find the reconciliation between the two views.

> — *Norbert Wiener*

Do not undertake a scientific career in quest of fame or money. There are easier and better ways to reach them. Undertake it only if nothing else will satisfy you; for nothing is probably what you will receive. Your reward will be the widening of the horizon as you climb. And if you achieve that reward you will ask no other.

> — *Cecilia Payne Gaposchkin*
> In *An Autobiography and Other Recollections*, 1984

As an adolescent I aspired to lasting fame, I craved factual certainty, and I thirsted for a meaningful vision of human life—so I became a scientist. This is like becoming an archbishop so you can meet girls.

> — *Matt Cartmill*

You too can be a toxicologist in two easy lessons, each of ten years.

> — *Arnold Lehman*, 1955

Divisions in Science

Specialization is for insects.

> — *Robert Heinlein*
> In *Time Enough for Love*, 1973

The field cannot well be seen from within the field.

> — *Ralph Waldo Emerson*

Where the telescope ends, the microscope begins. Which of the two has the grander view?

> — *Victor Hugo*
> In *Les Miserables*, 1862
>
> *Perhaps this breakdown is still true in modern physics which tends to be divided between the study of the heavens and the study of atomic particles. Or has the speaker simply forgotten about all of the things that can be seen with the naked eye that lie between the extremes?*

It is my intent to beget a good understanding between the chymists
and the mechanical philosophers who have hitherto been too little
acquainted with one another's learning.

> — *Robert Boyle*
> In *The Sceptical Chymist*, 1661
> *Thus lie the origins of Chemical Engineering?*

If it is granted that biodiversity is at high risk, what is to be done? The
solution will require cooperation among professions long separated by
academic and practical tradition.

> — *Edward O. Wilson*
> In *The Diversity of Life*, 1992

We need more rather than fewer classifications, different classifications,
always new classifications, to meet new needs.

> — *J. O. Brew*

There are no sects in geometry.

> — *Voltaire*

Science and Art belong to the world as a whole and the barriers of
nationality vanish before them.

> — *Johann Goethe, 1813*

Most "scientists" are bottle washers and button sorters.

> — *Robert Heinlein*
> In *Time Enough for Love*, 1973

All science is either physics or stamp collecting.

> — *Ernest Rutherford*
> Quoted in *Rutherford at Manchester* by J.B. Birks
> *In Rutherford's time, the end of the 19th century, much of science was either
> based heavily on physics or consisted mostly of collecting specimens. This is
> less true now, a century later.*

The difference between science and the fuzzy subjects is that science
requires reasoning, while those other subjects merely require scholarship.

> — *Robert Heinlein*
> In *Time Enough for Love*, 1973

The art of medicine in Egypt is thus exercised: one physician is confined
to the study and management of one disease; there are of course a great
number who practice this art; some attend to the disorders of the eyes,
others to those of the head, some take care of the teeth, others are
conversant with all diseases of the bowels; whilst many attend to the
cure of maladies which are less conspicuous.

> — *Herodotus*
> In *Book II, Euterpe*, circa 450 B.C.E.
> *The nature of specialization is not a new phenomenon.*

Through limiting their vision by phyletic boundaries, primatologists
have too often tackled issues with which ornithologists were already
highly experienced.

> — *Patrick Bateson and Robert Hinde*
> In *Growing Points in Ethology*, 1976
> *Definitions: phyletic—taxonomic, i.e. limiting themselves to certain species;*
> *primatologists—people who study monkeys and apes; ornithologists—*
> *people who study birds.*

Engineers think that equations approximate the real world.
Physicists think that the real world approximates equations.
Mathematicians are unable to make the connection.

> — *Anonymous*

If it's green or wriggles, it's biology.
If it stinks, it's chemistry.
If it doesn't work, it's physics.
If it's useful, it's engineering.

> — *Anonymous*

Chapter 2
Mathematics

Mathematics is the gate and key to the sciences.
— *Roger Bacon*

Mathematics seems to endow one with something like a new sense.
— *Charles Darwin*

Like the crest of a peacock so is mathematics at the head of all knowledge.
— *Indian proverb*

The control of large numbers is possible, and like unto that of small numbers, if we subdivide them.
— *Sun Tzu*

It has been said that figures rule the world. Maybe. But I am sure that figures show us whether it is being ruled well or badly.
— *Johann Goethe*
Quoted in *Conversations with Goethe* by J. Eckermann

The latest authors, like the most ancient, strove to subordinate the phenomena of nature to the laws of mathematics.
— *Sir Isaac Newton*

Mathematics has the completely false reputation of yielding infallible conclusions. Its infallibility is nothing but identity. Two times two is not four, but is just two times two, and that is what we call four for short. But four is nothing new at all. And thus it goes on and on in its conclusions, except that in the higher formulas the identity fades out of sight.
— *Johann Goethe*
Quoted in *The World of Mathematics* by J. Newman

If we pass this bill which establishes a new and correct value of pi, the author offers our state without cost the use of this discovery and its free publication in our school textbooks, while everyone else must pay him a royalty.
— *Indiana legislature, February 1897*
Quoted in *A History of Pi* by Petr Beckmann
Sample of debate about a bill to redefine the value of pi to be 9.2376. The Indiana House of Legislatures unanimously passed the legislation. The Indiana Senate did not take up the bill. The actual value of pi is 3.14159…

Math is like love—a simple idea but it can get complicated.
— *R. Drabek*

The imaginary number is a fine and wonderful recourse of the divine
spirit, almost an amphibian between being and not being.
— *Gottfried Wilhelm Leibniz*

The Mean Value Theorem is the midwife of calculus—not very important
or glamorous by itself, but often helping to delivery other theorems that
are of major significance.
— *E. Purcell and D. Varberg*
In *Calculus with Analytic Geometry*, 1987
*The Mean Value Theorem essentially states that at some point on a
graph between endpoints, the tangent line is parallel to the chord
joining those endpoints.*

Mathematics is like checkers in being suitable for the young, not too
difficult, amusing, and without peril to the state.
— *Plato*
Perhaps mathematics was simpler at that time.

In mathematics you don't understand things. You just get used to them.
— *Johann von Neumann*
Quoted in *The Dancing Wu Li Masters* by G. Zukav

Mathematics is the cheapest science. Unlike physics or chemistry, it does
not require any expensive equipment. All one needs for mathematics is a
pencil and paper.
— *George Polya*
Quoted in *Mathematical People* by D. Albers and G. Alexanderson
And sometimes the mathematician wants a powerful computer.

The mathematician…must often have the uncomfortable feeling that his
paper and pencil surpass him in intelligence.
— *Ernst Mach*
In "The Economy of Science" in *The World of Mathematics*
edited by J.R. Newman, 1956

One cannot escape the feeling that these mathematical formulas have an independent existence and an intelligence of their own, that they are wiser than we are, wiser even than their discoverers, that we get more out of them than was originally put into them.

> — *Heinrich Hertz*
> Quoted in *Men of Mathematics* by E. T. Bell

It can be of no practical use to know that Pi is irrational, but if we can know, it surely would be intolerable not to know.

> — *E. C. Titchmarsh*
> *An irrational number is one that cannot be expressed as a ratio of two integers.*

I am ill at these numbers.

> — *William Shakespeare*
> In *Hamlet*, 1601

Mathematics make them [people] sad.

> — *Martin Luther*

I tell them that if they will occupy themselves with the study of mathematics they will find in it the best remedy against the lusts of the flesh.

> — *Thomas Mann*
> In *The Magic Mountain*, 1927

The traditional mathematics professor of the popular legend is absent-minded. He usually appears in public with a lost umbrella in each hand. He prefers to face the blackboard and to turn his back to the class. He writes *a*; he says *b*; he means *c*; but it should be *d*. Some of his sayings are handed down from generation to generation.

"In order to solve this differential equation you look at it till a solution occurs to you."

"The principle is so perfectly general that no particular application of it is possible."

"Geometry is the science of correct reasoning on incorrect figures."

"My method to overcome a difficulty is to go around it."

"What is the difference between method and device? A method is a device which you used twice."

> — *George Polya*
> In *How to Solve It*, 1945

A polar bear is a rectangular bear after a coordinate transform.
> — *Anonymous*

A good mathematical joke is better, and better mathematics, than a dozen mediocre papers.
> — *J.E. Littlewood*
> In *A Mathematician's Miscellany*, 1953

Arithmetic is being able to count up to twenty without taking off your shoes.
> — *attributed to Mickey Mouse*
> *Please note, however, that Mickey Mouse only has four fingers on each hand. Then again, perhaps he was counting in base 2?*

Proofs

Mathematical proofs, like diamonds, are hard and clear, and will be touched with nothing but strict reasoning.
> — *John Locke*
> Quoted in *Elementary Number Theory* by D. Burton

Proof is the idol before whom the pure mathematician tortures himself.
> — *Sir Arthur Eddington*

We never make assertions… We do not tell—we *show*. We do not claim—we *prove*.
> — *Ayn Rand*
> In *Atlas Shrugged*, 1957

We often hear that mathematics consists mainly of "proving theorems." Is a writer's job mainly that of "writing sentences?"
> — *Gain-Carlo Rota*
> In preface to *The Mathematical Experience* by P. Davis and R. Hersh, 1981

It is the first duty of a hypothesis to be intelligible.
> — *Thomas Henry Huxley*

Jurij Vega (1754–1802) is best known for his writngs on logarithms and trigonometric functions. He is featured on the 1992 Slovenian 50 Tolar note. The left hand area of the banknote features a drawing from Vega's *Treatise on the Sphere.*

A proof tells us where to concentrate our doubts.
> *— Morris Kline*

"Obvious" is the most dangerous word in mathematics.
> *— Eric Temple Bell*

Mathematics is a game played according to certain simple rules with meaningless marks on paper.
> *— David Hilbert*

Mathematics consists of proving the most obvious thing in the least obvious way.
> *— George Polya*

Though this be madness, yet there is method in it.
> *— William Shakespeare*
> In *Hamlet*, 1601

It is a good morning exercise for a research scientist to discard a pet hypothesis every day before breakfast.
> *— Konrad Lorenz*
> In *On Aggression*, 1966

Mathematics is not a careful march down a well-cleared highway, but a journey into a strange wilderness, where the explorers often get lost. Rigour should be a signal to the historian that the maps have been made, and the real explorers have gone elsewhere.
> *— W.S. Anglin*
> In "Mathematics and History" in *Mathematical Intelligencer*

Young men should prove theorems, old men should write books.
— *Godfrey Hardy*

Q: Why did the chicken cross the road?
A: Pierre de Fermat: I just don't have room here to give the
full explanation.
— *Anonymous*

> *Fermat was a 17th century mathematician most famous for "Fermat's last theorem," an intriguing mathematical theorem written in the margin of his notebook shortly before he died. After the mathematical formula, he added the words "I have discovered a truly remarkable proof which this margin is too small to contain." After more than 3 centuries, a proof for the theorem was finally found in 1994.*

Statistics

Statistics are the heart of democracy.
— *Simeon Strunsky*
> In "Topics of the Times," November 30, 1944

Life is a school of probability.
— *Walter Bagehot*
> Quoted in *The World of Mathematics* by J. Newman

Experimentalists think that it is a mathematical theorem while the
mathematicians believe it to be an experimental fact.
— *Gabriel Lippman*
> Quoted in *On Growth and Form* by D'Arcy Thompson
> *Speaking to Poincaré about the Gaussian distribution (a.k.a. normal distribution or bell curve).*

If your experiment needs statistics, you ought to have done a
better experiment.
— *Ernest Rutherford*
> Quoted in *The Mathematical Approach to Biology and Medicine* by N. Bailey
> *No, you apply statistics in order to better understand the problem.*

Facts speak louder than statistics.
— *Geoffrey Streatfield*

Probability has turned modern science into a truth casino.
— *Bart Kosko*
  In *Fuzzy Thinking: The New Science of Fuzzy Logic*, 1994

Chance governs all.
— *John Milton*

How dare we speak of the laws of chance? Is not chance the antithesis of all law?
— *Bertrand Russell*

Of course, the entire effort is to put oneself outside the ordinary range of what are called statistics.
— *Stephen Spender*
  Quoted in *Chaos: Making a New Science* by James Gleick

The roulette wheel has neither conscience nor memory.
— *Joseph Bertrand*

Statistics: the mathematical theory of ignorance.
— *Morris Kline*

Statistics are like alienists—they will testify for either side.
— *Fiorello La Guardia*
  An alienist is a physician who diagnoses mental disorders for legal trials.

There are three kinds of lies—lies, damn lies, and statistics.
— *attributed to Mark Twain*
  Twain says that he obtained the quote from Benjamin Disraeli. Unfortunately Disraeli is not known to have said this. It could have come from a member of the Royal Statistical Society who said in 1896, "We may quote to one another with a chuckle the words of the Wise Statesman, lies, damned lies, statistics...."

The country is hungry for information; everything of a statistical character, or even a statistical appearance, is taken up with an eagerness that is almost pathetic; the community has not yet learned to be half skeptical and critical enough in respect to such statements.

> — *General Frances A. Walker*
> *Superintendent of the 1870 census.*

It is difficult to understand why statisticians commonly limit their enquiries to averages, and do not revel in more comprehensive views. Their souls seem as dull to the charm of variety as that of the native of one of our flat English counties, whose retrospect of Switzerland was that, if its mountains could be thrown into its lakes, two nuisances would be got rid of at once.

> — *Sir Francis Galton*

A weaker man might be moved to re-examine his faith, if in nothing else at least in the law of probability.

> — *Tom Stoppard*
> In *Rosencrantz and Guildenstern are Dead*, 1976
> *Rosencrantz and Guildenstern have been flipping a coin that has turned heads 85 times in a row: statistically improbable, but not impossible.*

The law of averages, if I have got this right, means that if six monkeys were thrown up in the air for long enough they would land on their tails about as often as they would land on their heads.

> — *Tom Stoppard*
> In *Rosencrantz and Guildenstern are Dead*, 1976
> *The playwright is making fun of the saying that if enough monkeys were placed in front of typewriters and given enough time, then they would type* Hamlet. *The play is, itself, a satire based upon* Hamlet.

Her statistics were more than a study, they were indeed her religion…. Florence Nightingale believed … [that] we must study statistics, for these are the measure of His purpose. Thus the study of statistics was for her a religious duty.

> — *Karl Pearson*
> In *The Life, Letters and Labours for Francis Galton*, 1924
> *Florence Nightingale founded the modern profession of nursing and used statistics to validate her hospital reforms.*

Theory and Modeling

The purpose of models is not to fit the data but to sharpen the questions.
—*Samuel Karlin*
In the 11ᵗʰ R.A. Fisher Memorial Lecture at the Royal Society, April 1983

He who loves practice without theory is like the sailor who boards ship without a rudder and compass and never knows where he may cast.
— *Leonardo da Vinci*

And yet a relation appears, A small relation expanding like the shade of a cloud on sand, a shape on the side of a hill.
— *Wallace Stevens*
In "Connoisseur of Chaos" in *The Collected Poems of Wallace Stevens*, 1990

Science only begins for man from the moment when his mind lays hold of matter—when he tries to subject the mass accumulated by experience to rational combinations.
— *Alexander von Humboldt*
In *Cosmos, a Sketch of the Universe*, 1848

The great tragedy of Science—the slaying of a beautiful hypothesis by an ugly fact.
—*Thomas Huxley*
In *Collected Essays*, "Biogenesis and Abiogenesis"

A theory has only the alternative of being right or wrong. A model has a third possibility: it may be right, but irrelevant.
— *Manfred Eigen*
Quoted in *The Physicist's Conception of Nature* by Jagdish Mehra

The sciences do not try to explain, they hardly even try to interpret, they mainly make models. By a model is meant a mathematical construct which, with the addition of certain verbal interpretations, describes observed phenomena. The justification of such a mathematical construct is solely and precisely that it is expected to work.
—*Johann von Neumann*

The best material model of a cat is another, or preferably the same, cat.
— *A. Rosenblueth*
In *Philosophy of Science*, 1945
Although it might be a "better" model, it certainly would not be a very useful model, for then how could one extrapolate to greater knowledge?

No one believes the results of the computational modeler except the modeler, for only he understands the premises. No one doubts the experimenter's results except the experimenter, for only he knows his mistakes.
— *Anonymous*

Theory attracts practice as the magnet attracts iron.
— *Karl Friedrich Gauss*

Don't ask what it means, but rather how it is used.
— *Ludwig Wittgenstein*

All science has one aim, namely, to find a theory of nature.
— *Ralph Waldo Emerson*
In *Nature*, 1836

A model is a theoretical construct that provides an explanatory idea.
— *Edward Redish*
In a talk titled "Making Sense of What Happens in Physics Classes: Analyzing Student Learning" given at the American Physical Society on March 24, 1999

The chief aim of all investigations of the external world should be to discover the rational order and harmony which has been imposed on it by God and which He revealed to us in the language of mathematics.
— *Johannes Kepler*

Physicists like to think that all you have to do is say, these are the conditions, now what happens next?
— *Richard Feynman*
Nobel Prize winning physicist reflecting upon chaos theory which suggests that the outcome of some phenomena cannot be predicted based upon approximate measures of their initial conditions.

It is impossible to trap modern physics into predicting anything with perfect determinism because it deals with probabilities from the outset.
> — *Sir Arthur Eddington*
> Quoted in *The World of Mathematics* by J. Newman

Prediction is difficult, especially the future.
> — *Niels Bohr*

Comprehensive models are built, if at all, by many hands over many decades.
> — *James Doran*

The universe is simple; it's the explanation that's complex.
> — *Anonymous*

Euclid taught me that without assumptions there is no proof. Therefore, in any argument, examine the assumptions.
> — *Eric Temple Bell*
> Quoted in *Return to Mathematical Circles* by H. Eves

Accurate reckoning—the entrance into the knowledge of all existing things and all obscure secrets.
> — *Ahmes the Scribe, 17th century B.C.E.*
> Quoted in *A History of Pi* by Petr Beckmann

The science of mathematics presents the most brilliant example of how pure reason may successfully enlarge its domain without the aid of experience.
> — *Immanuel Kant*
> Quoted in *The Mathematical Intelligencer*

Whenever anyone says, "theoretically," they really mean, "not really."
> — *Dave Parnas*

When I am working on a problem I never think about beauty. I only think about how to solve the problem. But when I have finished, if the solution is not beautiful, I know it is wrong.
> — *Buckminster Fuller*

We should be careful to get out of an experience only the wisdom that is in it—and stop there; lest we be like the cat that sits down on a hot stove lid. She will never sit down on a hot stove lid again—and that is well; but also she will never sit down on a cold one any more.
— *Mark Twain*

It is the mark of an educated mind to rest satisfied with the degree of precision which the nature of the subject admits and not to seek exactness where only an approximation is possible.
— *Aristotle*

It is a mistake to believe that a science consists of nothing but conclusively proven propositions, and it is unjust to demand that it should. It is a demand only made by those who feel a craving for authority in some form and a need to replace the religious catechism by something else, even if it be a scientific one.
— *Sigmund Freud*

One of the endlessly alluring aspects of mathematics is that its thorniest paradoxes have a way of blooming into beautiful theories.
— *Philip Davis*
In *Scientific American*, September 1964
This is true in all science.

Equations are more important to me, because politics is for the present, but an equation is something for eternity.
— *Albert Einstein*
Quoted in *A Brief History of Time* by Stephen Hawking

It's an experience like no other experience I can describe, the best thing that can happen to a scientist, realizing that something that's happened in his or her mind exactly corresponds to something that happens in nature. It's startling every time it occurs. One is surprised that a construct of one's own mind can actually be realized in the honest to goodness world out there. A great shock, and a great, great joy.
— *Leo Kadanoff*

Math in Science

No human investigation can be called real science if it cannot be demonstrated mathematically.
— *Leonardo da Vinci*

If it can't be expressed in figures, it is not science; it is opinion.
— *Robert Heinlein*
In *Time Enough for Love*, 1973

There is something fascinating about science. One gets such wholesale returns of conjecture out of such a trifling investment of fact.
— *Mark Twain*

The more progress physical sciences make, the more they tend to enter the domain of mathematics, which is kind of centre to which they all converge. We may even judge of the degree of perfection to which a science has arrived by the facility with which it may be submitted to calculation.
— *Adolphe Quetelet*

Abu Nasr Al-Farabi (870–950) appears on the 1 Tenge note from Kazakhstan. A scholar in many areas, including philosophy, linguistics, lobis, and music, he also wrote about the nature of science and argued for the existence of the vacuum. The reverse of the note features his mathematical drawings and equations.

[The universe] cannot be read until we have learnt the language and become familiar with the characters in which it is written. It is written in mathematical language, and the letters are triangles, circles and other geometrical figures, without which means it is humanly impossible to comprehend a single word.
— *Galileo Galilei*
In *Opere II Saggiatore*

Mathematics may be compared to a mill of exquisite workmanship, which grinds your stuff of any degree of fineness; but, nevertheless, what you get out depends on what you put in; and as the grandest mill in the world will not extract wheat flour from peascods, so pages of formulae will not get a definite result out of loose data.
— *Thomas Henry Huxley*
In the Quarterly Journal of the Geological Society, 1869

Data without generalization is just gossip.
— *Robert Pirsig*

One should always generalize.
— *Carl Jacobi*
Quoted in *The Mathematical Experience* by P. Davis and R. Hersh

To generalize is to be an idiot.
— *William Blake*

All generalizations are dangerous, even this one.
— *Alexandre Dumas*
Quoted in *The Boston Globe*

Balance Between Theory and Reality

In theory, there is no difference between theory and practice. But, in practice, there is.
—*Jan L.A. van de Snepscheut*

Textbooks and Heaven only are Ideal.
—*John Updike*
In *Dance of the Solids*

We think in generalities, but we live in details.
—*Alfred North Whitehead*

The brightest flashes in the world of thought are incomplete until they
have been proved to have their counterparts in the world of fact.
—*John Tyndall*
In *Fragments of Science*, 1863

The facts will promptly blunt his ardor.
—*Caecilius Statius*
In *The Changeling*

If the facts don't fit the theory, change the facts.
—*Albert Einstein*

It is fatal as it is cowardly to blame facts because they are not to
our taste.
—*John Tyndall*
In an address at Belfast, August 19, 1874

Experimental confirmation of a prediction is merely a measurement. An
experiment disproving a prediction is a discovery.
—*Enrico Fermi*

There are no such things as applied sciences, only applications of science.
—*Louis Pasteur*
In a speech on September 11, 1872

Although this may seem a paradox, all exact science is dominated by the
idea of approximation.
—*Bertrand Russell*

Science may be described as the art of systematic over-simplification.
—*Karl Popper, August 1982*

Never worry about theory as long as the machinery does what it's
supposed to do.
> — *attributed to Robert A. Heinlein*

A modern mathematical proof is not very different from a modern
machine, or a modern test setup; the simple fundamental principles are
hidden and almost invisible under a mass of technical details.
> — *Hermann Weyl*
> In *Unterrichtsblatter fur Mathematick and Naturwissenschaften*,
> 1932. Quoted in and translated by A. Shenitzer in *The American
> Mathematical Monthly*

The philosophers have only interpreted the world; the thing, however, is
to change it.
> — *Karl Marx*
> *All of the theories in the world amount to little if there is no ability to apply
> that knowledge to make the world a better place.*

The discovery in 1846 of the planet Neptune was a dramatic and
spectacular achievement of mathematical astronomy. The very existence
of this new member of the solar system, and its exact location, were
demonstrated with pencil and paper; there was left to observers only
the routine task of pointing their telescopes at the spot the mathematicians had marked.
> — *James R. Newman*
> In *The World of Mathematics*, 1956

Never does nature say one thing and wisdom another.
> — *Juvenal, ca. 100 C.E.*

Truth is stranger than fiction, but it is because fiction is obliged to stick
to possibilities; Truth isn't.
> — *Mark Twain*
> In *Pudd'nhead Wilson's New Calendar*, 1897

If the doors of perception were cleansed, everything would be seen as it
is, infinite.
> — *William Blake*

I didn't think; I experimented.
— *Wilhelm Roentgen*

That's not an experiment you have there; that's an experience.
— *Sir Ronald Fisher*

It is a capital mistake to theorize before one has data. Insensibly one begins to twist facts to suit theories instead of theories to suit facts.
— *Sir Arthur Conan Doyle*
However, one must have a theory before one knows where to look for the facts.

It is more important to have beauty in one's equations than to have them fit experiment. If Schroedinger had been more confident of his work, he could have published it months earlier, and he could have published a more accurate equation. If there is not complete agreement between the results of one's work and experiment, one should not allow oneself to be too discouraged, because the discrepancy... will get cleared up with further development of the theory.
— *Paul Dirac*
In *Scientific American*, May 1963

In solving a problem of this sort, the grand thing is to be able to reason backwards... Most people, if you describe a train of events to them, will tell you what the result would be... There are few people, however, who, if you told them a result, would be able to evolve from their own inner consciousness what the steps were which led up to that result.
— *Sir Arthur Conan Doyle*
The most interesting experiments tend to provide an unexpected result whose origins must be reasoned backwards.

Intricate relationships between the Mississippi and its tributaries do not always lend themselves to computer analysis. A computer is only as good as the data you input. There usually comes a time, in these situations, when you feel a lot better if you can put the problem on the model and get an empirical answer.
— *Major General Charles Noble, 1973*
Quoted in *A History of the Waterways Experiment Station*
In reference to using physical models to assist with the understanding of the 1973 Mississippi flood.

Now, if I wanted to be one of those ponderous scientific people, and "let on" to prove what had occurred in the remote past by what had occurred in a given time in the recent past, or what will occur in the far future by what has occurred in late years, what an opportunity is here! Geology never had such a chance, nor such exact data to argue from! Please observe:

In the space of 176 years, the Lower Mississippi has shortened itself 242 miles. That is an average of a trifle over one mile and a third per year. Therefore, any calm person, who is not blind or idiotic, can see that in the old Oolitic Silurian Period, just upwards to a million years ago next November, the Lower Mississippi River was upwards to 1,300,000 miles long, and stuck out over the Gulf of Mexico like a fishing rod. And by the same token, any person can see that 742 years from now the Lower Mississippi will only be a mile and three-quarters long, and Cairo and New Orleans will have joined their streets together, and be plodding comfortably along under a single mayor and a mutual board of aldermen. There is something fascinating about science. One gets such wholesome returns of conjecture out of such a trifling investment in fact.

> — *Mark Twain*
> In *Life on the Mississippi*, 1944. Quoted in *Streambank Stabilization Handbook.*
> *Moral: Beware making broad generalizations from limited data. Extrapolation beyond the data is very dangerous!*

I have seen so much of the danger arising from presenting results or rules involving variable coefficients in the form of algebraic formulas which the hurried or careless worker may use far beyond the limit of the experimental determination that I present the results mainly in the form of plotted curves which cannot be thus misused and which clearly show the margin of uncertainty and the limitations of the data.

> — *John Freeman*
> In *Experiments upon the Flow of Water in Pipes and Pipe Fittings*, 1941

Relevance of Math

No knowledge can be certain if it is not based upon mathematics or upon some other knowledge which is itself based upon the mathematical sciences.

> — *Leonardo da Vinci*

For the things of this world cannot be made known without a knowledge of mathematics.
— *Roger Bacon*
In *Opus Majus*, 1267

Mathematics is the science which draws necessary conclusions.
— *Benjamin Pierce*
In a memoir read before the National Academy of Sciences in Washington, 1870

Pure mathematics, may it never be of any use to anyone.
— *Henry John Stephen Smith*
Spoken as a toast.

Mathematics is the tool specially suited for dealing with abstract concepts of any kind and there is not limit to its power in this field.
— *Paul Dirac*

I admit that mathematical science is a good thing. But excessive devotion to it is a bad thing.
— *Aldous Huxley*
In an interview with J.W.N. Sullivan in *Contemporary Mind*, 1934

The science of calculation also is indispensable as far as the extraction of the square and cube roots: Algebra as far as the quadratic equation and the use of logarithms are often of value in ordinary cases: but all beyond these is but a luxury: a delicious luxury indeed: but not to be indulged in by one who is to have a profession to follow for his subsistence.
— *Thomas Jefferson*
Quoted in "The Encouragement of Science" by J. Robert Oppenheimer in *The Armchair Science Reader* edited by I. Gordon and S. Sorkin

Mathematical rigor is like clothing; in its style it ought to suit the occasion, and it diminishes comfort and restrains freedom of movement if it is either too loose or too tight.
— *G.F. Simmons*

Whoever despises the high wisdom of mathematics nourishes himself on delusion.
— *Leonardo da Vinci*

One of the endearing things about mathematicians is the extent to which they will go to avoid doing any real work.
— *Matthew Pordage*

God forbid that Truth should be confined to Mathematical Demonstration!
— *William Blake*
In *Notes on Reynold's Discourses,* 1808

What would life be without arithmetic, but a scene of horrors.
— *Sydney Smith*

Mathematics takes us into the region of absolute necessity, to which not only the actual world, but every possible world, must conform.
— *Bertrand Russell*
In *The Study of Mathematics*

How can it be that mathematics, being after all a product of human thought independent of experience, is so admirably adapted to the objects of reality?
— *Albert Einstein*

The Force of Numbers, which can be successfully applied, even to those things, which one would imagine are subject to no Rules.
— *Christiaan Huygens*
In *De Ratiociniis in Ludo Aleae,* 1657
The title translates to **On Computations in the Game of Dice.** *The book describes Blaise Pascal's work in statistics and probability.*

There are very few things which we know which are not capable of being reduc'd to a Mathematical Reasoning, and when they cannot, it's a sign our knowledge of them is very small and confused.
— *Christiaan Huygens*
In *De Ratiociniis in Ludo Aleae,* 1657

For a physicist, mathematics is not just a tool by means of which phenomena can be calculated, it is the main source of concepts and principles by means of which new theories can be created.
— *Freeman Dyson*
In *Mathematics in the Physical Sciences*

Mathematics is not only real, but it is the only reality… The entire universe is made out of particles [electrons, protons, etc.]. Now what are the particles made out of? They're not made out of anything. The only thing you can say about the reality of an electron is to cite its mathematical properties. There's a sense in which matter has completely dissolved and what is left is just a mathematical structure.
> — *Martin Gardner*
> In *Focus—The Newsletter of the Mathematical Association of America,* December, 1994

Mathematics began to seem too much like puzzle solving. Physics is puzzle solving, too, but puzzles created by nature, not by the mind of man.
> — *Maria Goeppert-Mayer*
> Quoted in *Maria Goeppert-Mayer, A Life of One's Own* by J. Dash

As far as the laws of mathematics refer to reality, they are not certain, and as far as they are certain, they do not refer to reality.
> — *Albert Einstein*

Mathematics is an interesting intellectual sport but it should not be allowed to stand in the way of obtaining sensible information about physical processes.
> — *Richard Hamming*

The most painful thing about mathematics is how far away you are from being able to use it after you have learned it.
> — *James R. Newman*
> In *The World of Mathematics,* 1956

It can be shown that a mathematical web of some kind can be woven about any universe containing several objects. The fact that our universe lends itself to mathematical treatment is not a fact of any great philosophical significance.
> — *Bertrand Russell*

I am doing mathematics. But it is relevant to what is around us. That is nature, too.
> — *Albert Libchaber*

Without mathematics one cannot fathom the depths of philosophy;
Without philosophy one cannot fathom the depths of mathematics;
without the two, one cannot fathom anything.
> — *Bordas-Demoulins*

The most distinctive characteristic which differentiates mathematics
from the various branches of empirical science, and which accounts for
its fame as the queen of sciences, is no doubt the peculiar certainty and
necessity of its results.
> — *Carl Hempel*
> In "Geometry and Empirical Science" in *The World of Mathematics*, 1956

The mathematical facts worthy of being studied are those which, by their
analogy with other facts, are capable of leading us to the knowledge of a
physical law. They reveal the kinship between other facts, long known,
but wrongly believed to be strangers to one another.
> — *Jules Henri Poincaré*

Mathematicians have long since regarded it as demeaning to work on
problems related to elementary geometry in two or three dimensions,
in spite of the fact that it is precisely this sort of mathematics which is
of practical value.
> — *Branko Grunbaum and G. C. Shephard*
> In the *Handbook of Applicable Mathematics*

All mathematical laws which we find in Nature are always suspect to
me, in spite of their beauty. They give me no pleasure. They are merely
auxiliaries. At close range it is all not true.
> — *George Christoph Lichtenberg*
> In *Lichtenberg* by J. P. Stern

This trend [emphasizing applied mathematics over pure mathematics]
will take the queen of the sciences into the quean of the sciences.
> — *L.M. Passano*
> A quean is a prostitute.

The first law of Engineering Mathematics: All infinite series converge,
and moreover converge to the first term.
> — *Anonymous*

Math as Art

I am interested in mathematics only as a creative art.
> — *Godfrey Hardy*
> In *A Mathematician's Apology*, 1941

The mathematical sciences particularly exhibit order, symmetry, and limitation; and these are the greatest forms of the beautiful.
> —*Aristotle*
> In *Metaphysica*, 60 B.C.E.

Mathematics possesses not only truth, but supreme beauty—a beauty cold and austere, like that of sculpture, without appeal to any part of our weaker nature, sublimely pure, and capable of a stern perfection such as only the greatest art can show.
> — *Bertrand Russell*
> In *The Study of Mathematics*

Abstractness, sometimes hurled as a reproach at mathematics, is its chief glory and its surest title to practical usefulness. It is also the source of such beauty as may spring from mathematics.
> — *Eric Temple Bell*

Creative mathematicians now, as in the past, are inspired by the art of mathematics rather than by any prospect of ultimate usefulness.
> — *Eric Temple Bell*

The mathematician's best work is art, a high perfect art, as daring as the most secret dreams of imagination, clear and limpid. Mathematical genius and artistic genius touch one another.
> — *Gosta Mittag Leffler*

Pure mathematics is on the whole distinctly more useful than applied. For what is useful above all is technique, and mathematical technique is taught mainly through pure mathematics.
> — *Godfrey Hardy*

Mathematicians do not study objects, but relations between objects. Thus, they are free to replace some objects by others so long as the relations remain unchanged. Content to them is irrelevant; they are interested in form only.
— *Jules Henri Poincaré*

The four branches of arithmetic—ambition, distraction, uglification and derision.
— *Lewis Carroll*
    In *Alice in Wonderland*, 1865

In my opinion, a mathematician, in so far as he is a mathematician, need not preoccupy himself with philosophy—an opinion, moreover, which has been expressed by many philosophers.
— *Henri Lebesgue*
    In *Scientific American*, September 1964

Mathematicians

The mathematician has reached the highest rung on the ladder of human thought.
— *Havelock Ellis*
    In *The Dance of Life*

He who can properly define and divide is to be considered a god.
— *Plato*

The mind that constantly applies itself to geometry is not likely to fall into error. In this convenient way, the person who knows geometry acquires intelligence.
— *Ibn Khaldun*
    In *The Muqaddimah. An Introduction to History*, 1375

A Mathematician is a machine for turning coffee into theorems.
— *Paul Erdos*

All great theorems were discovered after midnight.
— *Adrian Mathesis*

To be a scholar of mathematics you must be born with talent, insight, concentration, taste, luck, drive and the ability to visualize and guess.
— *Paul Halmos*
In *I Want to be a Mathematician*, 1985

It would be better for the true physics if there were no mathematicians on earth.
— *Daniel Bernoulli*

I have hardly ever known a mathematician who was capable of reasoning.
— *Plato*
In The Republic, circa 390 B.C.E.

You know we all became mathematicians for the same reason: we were lazy.
— *Max Rosenlicht, 1949*

Mathematics is not yet capable of coping with the naivete of the mathematician himself.
— *Abraham Kaplan*
In *Sociology Learns the Language of Mathematics*

The good Christian should beware of mathematicians, and all those who make empty prophecies. The danger already exists that the mathematicians have made a covenant with the devil to darken the spirit and to confine man in the bonds of Hell.
— *Saint Augustine*
In *DeGenesi ad Litteram, Book II*
During the 4th century, mathematicians were often equated with astrologers.

I love only nature, and I hate mathematicians.
— *Richard Feynman*

Mathematicians boast of their exacting achievements, but in reality they are absorbed in mental acrobatics and contribute nothing to society.
— *Sorai Ogyu*
Quoted in the *Complete Works on Japan's Philosophical Thought*

With me everything turns into mathematics.
— *René Descartes*

In many cases, mathematics is an escape from reality. The mathematician finds his own monastic niche and happiness in pursuits that are disconnected from external affairs… In their unhappiness over the events of this world, some immerse themselves in a kind of self-sufficiency in mathematics.
— *Stanislaw Ulam*
In *Adventures of a Mathematician*, 1976

Life is good for only two things, discovering mathematics and teaching mathematics.
— *Simeon Poisson*
Quoted in *Mathematics Magazine*
It is a good that all of the world's farmers do not share this belief.

To state a theorem and then to show examples of it is literally to teach backwards.
— *E. Kim Nebeuts*

A topologist is one who doesn't know the difference between a doughnut and a coffee cup.
— *John Kelley*
Quoted in *Mathematical Maxims and Minims* by N. Rose
Topology is the study of geometric forms which remain invariant during transformations such as bending and stretching.

Number theorists are like lotus-eaters—having once tasted of this food they can never give it up.
— *Leopold Kronecker*
Described in Homer's The Odyssey, *Lotus-eaters live in a state of languorous forgetfulness from eating lotus fruit.*

Anyone who cannot cope with mathematics is not fully human. At best he is a tolerable subhuman who has learned to wear shoes, bathe, and not make messes in the house.

> — *Robert A. Heinlein*
> In *Time Enough for Love, 1973*

I'm sorry to say that the subject I most disliked was mathematics. I have thought about it. I think the reason was that mathematics leaves no room for argument. If you made a mistake, that was all there was to it.

> — *Malcolm X*

Now I feel as if I should succeed in doing something in mathematics, although I cannot see why it is so very important... The knowledge doesn't make life any sweeter or happier, does it?

> — *Helen Keller*
> In *The Story of My Life*, 1903

If I feel unhappy, I do mathematics to become happy. If I am happy, I do mathematics to keep happy.

> — *Alfred Renyi*
> Quoted in "The Work of Alfred Renyi" by P. Turan in *Matematikai Lapok*

The greatest unsolved theorem in mathematics is why some people are better at it than others.

> — *Adrian Mathesis*

An eloquent mathematician must, from the nature of things, ever remain as rare a phenomenon as a talking fish, and it is certain that the more anyone gives himself up to the study of oratorical effect the less will he find himself in a fit state to mathematicize.

> — *J.J. Sylvester*

Mathematics is written for mathematicians.

> — *Nicholaus Copernicus*
> In *De Revolutionibus*, 1530

I never could make out what those damned dots meant.

> — *Lord Randolph Churchill*
> Quoted in *Lord Randolph Churchill* by W.S. Churchill.
> *Speaking about decimal points.*

O King, for traveling over the country, there are royal roads and roads for common citizens; but in geometry there is one road for all.

> — *Menaecbmus*
> Quoted in *A History of Pi* by Petr Beckmann
> *In reply to his pupil, Alexander the Great, when asked for a shortcut to geometry.*

There is no royal road to geometry.

> — *Euclid*
> *In response to Ptolemy I, King of Egypt, who wished to study geometry without going over the thirteen parts of Euclid's* **Elements**. *The king said that a short cut would be agreeable.*

Out of nothing I have created a strange new universe.

> — *Janos Bolyai*
> *In reference to his creation of non-euclidean geometry in the mid 19th century.*

For God's sake, please give it up. Fear it no less than the sensual passion, because it, too, may take up all your time and deprive you of your health, peace of mind and happiness in life.

> — *Wolfgang Bolyai*
> Quoted in *The Mathematical Experience* by P. Davis and R. Hersh
> *Speaking to his son, Janos Bolyai, urging him to give up work on non-euclidean geometry.*

On Tuesday evening at Museum, at a ball in the gardens. The night was chill, I dropped too suddenly from Differential Calculus into ladies' society, and could not give myself freely to the change. After an hour's attempt so to do, I returned, cursing the mode of life I was pursuing; next morning I had already shaken hands, however, with Differential Calculus, and forgot the ladies.

> — *Thomas Archer Hirst*
> In his journal entry for August 10, 1851

She knew only that if she did or said thus-and-so, men would unerringly respond with the complimentary thus-and-so. It was like a mathematical formula and no more difficult, for mathematics was the one subject that had come easy to Scarlett in her schooldays.

> — *Margaret Mitchell*
> In *Gone With the Wind*, 1936

Leibniz never married; he had considered it at the age of fifty; but the person he had in mind asked for time to reflect. This gave Leibniz time to reflect, too, and so he never married.

> — *Bernard Learning Bovier Fontenelle*
> In *Eloge de learning Leibniz*

It is rare to find learned men who are clean, do not stink, and have a sense of humor.

> — *Duchess of Orleans*
> *Speaking about Gottfried Leibniz. Also attributed to being spoken by Charles Louis de Secondat Montesquieu.*

Being perpetually charmed by his familiar siren, that is, by geometry, he neglected to eat and drink and took no care of his person; that he was often carried by force to the baths, and when there he would trace geometrical figures in the ashes of the fire, and with his finger draws lines upon his body when it was anointed with oil, being in a state of great ecstasy and divinely possessed by his science.

> — *Plutarch*
> Quoted in *Calculus Gems* by G. Simmons
> *Speaking about Archimedes. It has also been said that "except for Archimedes mathematics is singularly naked of anecdotes." (Sylvia Warner in* **Mr. Fortune's Maggot**, *1927)*

Don't spoil my circle!

> — *Archimedes, 212 B.C.E.*
> *Spoken to an invading Roman soldier just before Archimedes was killed while working at his blackboard. Ernst Mach commented that, "Archimedes constructing his circle pays with his life for his defective biological adaptation to immediate circumstances."*

Now I will have less distraction.

> — *Leonhard Euler*
> Quoted in *In Mathematical Circles* by H. Eves
> *Spoken upon losing the use of his right eye.*

In my experience most mathematicians are intellectually lazy and especially dislike reading experimental papers. He [Rene Thomas] seemed to have very strong biological intuitions but unfortunately of negative sign.

> — *Francis Crick*
> In *What Mad Pursuit*, 1988

If you ask a mathematician what they do, you always get the same answer.
They think. They think about difficult and unusual problems. They do
not think about ordinary problems; they just write down the answers.
> — *M. Egrafov*
> In *Mathematics Magazine*, December 1992

I believe that mathematical reality lies outside us, that our function is
to discover or observe it, and that the theorems which we prove, and
which we describe grandiloquently as our "creations," are simply the
notes of our observations.
> — *Godfrey Hardy*
> In *A Mathematician's Apology*, 1941

I have never thought a boy should undertake abstruse or difficult
sciences, such as Mathematics in general, till fifteen years of age at
soonest. Before that time they are best employed in learning the
languages, which is merely a matter of memory.
> — *Thomas Jefferson*
> In a letter to Ralph Izard, 1788

No one really understood music unless he was a scientist, her father
had declared, and not just a scientist, either, oh, no, only the real ones,
the theoreticians, whose language was mathematics. She had not
understood mathematics until he had explained to her that it was the
symbolic language of relationships. "And relationships," he had told her,
"contained the essential meaning of life."
> — *Pearl S. Buck*
> In *The Goddess Abides*, 1972

No, I have been teaching it all my life, and I do not want to have my
ideas upset.
> — *Isaac Todhunter*
> *When asked whether he would like to see an experimental demonstration
> of conical refraction.*

Usually mathematicians have to shoot somebody to get this much publicity.
> — *Thomas R. Nicely*
> In the *Cincinnati Enquirer*, December 1994
> *About the attention that he received after finding the flaw in Intel's
> Pentium processor.*

In the company of friends, writers can discuss their books, economists the state of the economy, lawyers their latest cases, and businessmen their latest acquisitions, but mathematicians cannot discuss their mathematics at all. And the more profound their work, the less understandable it is.
— *Alfred Adler*
In "Reflections: Mathematics and Creativity" in *The New Yorker Magazine*, February 1972
The same can be said about theoreticians in many types of science.

I read in the proof sheets of Hardy on Ramanujan: "As someone said, each of the positive integers was one of his personal friends." My reaction was, "I wonder who said that; I wish I had." In the next proof-sheets I read (what now stands), "It was Littlewood who said…"
— *J.E. Littlewood*
In *A Mathematician's Miscellany*, 1953

A mathematician will assume everything except responsibility.
— *Anonymous*

Q: How do you tell that you are in the hands of the Mathematical Mafia?
A: They make you an offer that you can't understand.
— *Michael Cook*

I'm very good at integral and differential calculus. I know the scientific names of beings animalculous; In short, in matters vegetable, animal, and mineral, I am the very model of a modern Major-General.
— *W.S. Gilbert*
In *The Pirates of Penzance*, 1879
An animalcule is a microscopic animal.

Chapter 3
Engineering

On the sixth day God saw He couldn't do it all, so He created engineers.
— *Lois McMaster Bujold*
In *Falling Free*, 1988

A scientist can discover a new star, but he cannot make one. He would have to ask an engineer to do that.
— *Gordon L. Glegg, 1969*

A good scientist is a person with original ideas. A good engineer is a person who makes a design that works with as few original ideas as possible. There are no prima donnas in engineering.
— *Freeman Dyson*
In *Disturbing the Universe*, 1979

There is only one nature—the division into science and engineering is a human imposition, not a natural one. Indeed, the division is a human failure; it reflects our limited capacity to comprehend the whole.
— *Bill Wulf*

It is not uncommon for engineers to accept the reality of phenomena that are not yet understood, as it is very common for physicists to disbelieve the reality of phenomena that seem to contradict contemporary beliefs of physics.
— *H. Bauer*

The scientist describes what is; the engineer creates what never was.
— *Theodore von Kármán*

One man's "magic" is another man's engineering.
— *Robert Heinlein*
In *Time Enough for Love*, 1973

For 'tis the sport to have the engineer
Hoisted with his own petard.
— *William Shakespeare*
In *Hamlet*, 1601

A petard is an explosive device used by military engineers. The figurative meaning of this phrase is to have one's own weapons used against oneself.

Mechanical Engineering

Instrumental or mechanical science is the noblest and above all others, the most useful.
> — *Leonardo da Vinci*

Mechanics is the paradise of the mathematical sciences, because by means of it one comes to the fruits of mathematics.
> — *Leonardo da Vinci*
> In *Notebooks*
>
> *In other words, without application in mechanical engineering, mathematics has little purpose.*

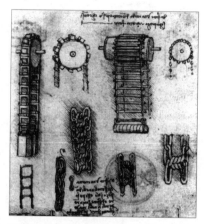

Sketches of chains and wheels with cogs for Leonardo da Vinci's conception of a bicycle (Madrid MS. I, f. 10r)

The description of right lines and circles, upon which geometry is founded, belongs to mechanics. Geometry does not teach us to draw these lines, but requires them to be drawn.
> — *Sir Isaac Newton*
> In *Principia Mathematica*, 1687
> *In other words, mechanics (and geometry) is dictated by application.*

Steel: A very remarkable, and most useful substance, prepared by heating the iron in contact with charcoal.
> — *George Fownes*
> In *Elementary Chemistry, Theoretical and Practical*, 1855.

There are few substances to which it yields in interest, when it is considered how very intimately the knowledge of the properties and uses of iron is connected with human civilization.
> — *George Fownes*
> In *Elementary Chemistry, Theoretical and Practical*, 1855

It is questionable if all the mechanical inventions yet made have lightened the day's toil of any human being.
— *John Stuart Mill*

Lo! Men have become the tools of their tools.
— *Henry David Thoreau*

When spider webs unite, they can tie up a lion.
— *Ethiopian proverb*
This is the essential principle behind composite manufacturing where each individual strand is weak, yet through load sharing, the resultant structure has great strength and high resiliency.

Moving parts in rubbing contact require lubrication to avoid excessive wear. Honorifics and formal politeness provide lubrication where people rub together.
— *Robert Heinlein*
In *Time Enough for Love*, 1973

If man has the intelligence to heat his house in the winter, why does he not cool in the summer? We go to the Arctic regions and heat our houses and live. We go down to the tropics and die.
— *Alexander Graham Bell, 1918*
Refrigeration techniques were invented in 1888. By 1901, the New York Stock Exchange was air-conditioned. However, what we know as air-conditioning was not in widespread use until the middle part of this century.

Give me where to stand, and I will move the world.
— *Archimedes*
Speaking about the lever whose principles he had just discovered.

If the only tool you have is a hammer, then every problem looks like a nail.
— *Anonymous*
And all you can do is whack the heck out of the problem.

Mr. Watson, come here, I want you.
> — *Alexander Graham Bell, March 10, 1876*
> *Spoken to his assistant as the first intelligible words transmitted by telephone.*

War will cease to be possible when all the world knows that tomorrow the most feeble of nations can supply itself immediately with a weapon which will render its host secure and its ports impregnable to the assaults of the united armadas of the world.
> — *Nikola Tesla, 1898*
> Quoted in *Scientific American*
> *Remarking about his remote control torpedo that uses wireless telegraphy for control.*

Anybody who goes anywhere without a roll of duct tape is a fool.
> — *Steve Mirsky*
> In *Scientific American*, July 1999
> *In an article discussing the Mir space station.*

Duct tape is like the force. It has a light side, and a dark side, and it holds the universe together.
> — *Carl Zwanzig*

Q: What is the difference between mechanical engineers and civil engineers?
A: Mechanical engineers build weapons, civil engineers build targets.
> — *Anonymous*

Engineering Design

To define it rudely but not inaptly, engineering… is the art of doing that well with one dollar, which any bungler can do with two after a fashion.
> — *Arthur Mellen Wellington*
> In *The Economic Theory of the Location of Railways*, 1900

Any intelligent fool can make things bigger, more complex and more violent. It takes a touch of genius and a lot of courage to move in the opposite direction.
> — *Albert Einstein*

Mankind cannot survive without technology. But unless technology becomes a true servant of man, the survival of mankind is in jeopardy. And if technology is to be the servant, then the engineer's paramount loyalty must be to society.
— *Victor Paschkis*

During their formative training, engineers are programmed with an irresistible drive to design systems and structures conservatively. Conservatism in design and specifications isn't a compulsion to waste effort and materials. It is the accumulation of centuries of experience.
— *H. W. Lewis*
In the *Introduction to Risk Analysis*, June 1999

Theoretical understanding, comprehension of practical and economic limitations, common sense, ability to do original and hard work—these are the requirements for a good design engineer, and they must be used in the approach to any design problem.
— *Max S. Peters, 1958*
Quoted in *Air Pollution Control—A Design Approach* by C. David Cooper and F. C. Alley

When we build, let us think that we build forever.
— *American proverb*

One cannot shingle the house before he builds it.
— *American proverb*

There was a tendency for some engineers to view public works as ends in themselves.
— *Michael Robinson*
Quoted in *Cadillac Desert* by Marc Reisner
Speaking about the Bureau of Reclamation.

I have always hated machinery, and the only machine I ever understood was a wheelbarrow, and that but imperfectly.
— *Eric Temple Bell*

Never use the T-square as a hammer.
Never sharpen a pencil over the drawing board.
Never leave the ink bottle uncorked.
Never use cheap materials of any kind.

— *Thomas French*
From a "Page of Cautions" in *Manual of Engineering Drawing*, 1924

Aeronautical Engineering

You cannot fly like an eagle with the wings of a wren.

— *William Henry Hudson*
In *Afoot in England*, 1909
Aerodynamic scaling would also tell the same information.

Flying without feathers is not easy; my wings have no feathers.

— *Titus Maccius Plautus*
In *Paenulus*, 220 B.C.E.
Original is "Sine pennis volare hau facilest: meae alea pennas non habent."

Oh, that I had wings like a dove, for then would I fly away, and be at rest.

— *Psalms 55:6*

Heavier-than-air flying machines are impossible.

— *William Thomson Kelvin, 1895*
President of the British Royal Society.

What can you conceive more silly and extravagant than to suppose a man racking his brains, and studying night and day how to fly?

— *William Law*
In *A Serious Call to a Devout and Holly Life XI*, 1728

Flying. Whatever any other organism has been able to do man should surely be able to do also, though he may go a different way about it.

— *Samuel Butler*

When once you have tasted flight, you will forever walk the earth with your eyes turned skyward, for there you have been, and there you will always long to return.

> — *Leonardo da Vinci*

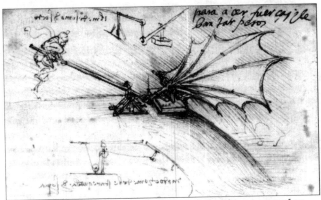

Leonardo da Vinci's sketch of a device to test the lifting power of an artificial wing. (Paris MS. B, f. 88)

The genius of Leonardo da Vinci imagined a flying machine, but it took the methodical application of science by those two American bicycle mechanics to create it.

> — *Bill Gates*
> Quoted in *Time Magazine*, 1999

The airplane stays up because it doesn't have the time to fall.

> — *Orville Wright*
> *One of the inventors of the airplane explaining the principles of powered flight.*

I confess that in 1901, I said to my brother Orville that man would not fly for fifty years ... Ever since, I have distrusted myself and avoided all predictions.

> — *Wilbur Wright, 1908*
> *Orville and Wilbur Wright flew the world's first airplane in 1903.*

I look with amazement upon our audacity in attempting flights with a new and untried machine.

> — *Orville Wright, 1914*
> Quoted in *Time Magazine*
> *Speaking about his 1903 flight of the first airplane.*

Aeronautics was neither an industry nor a science. It was a miracle.
> — *Igor Sikorsky*
>> *Sikorsky made the first American helicopter in 1939, three years after Heinrich Focke of Germany made the world's first helicopter.*

Science, freedom, beauty, adventure: What more could you ask of life? Aviation combined all the elements I loved.
> — *Charles Lindbergh*

There are no practical alternatives to air transportation.
> — *Daniel S. Goldin, March 20, 1997*

My children in their lifetime will see aeronautics become the greatest and principle means of national defense and rapid transportation all over the world and possibly beyond our world into interstellar space.
> — *Brigadier General Billy Mitchell*
>> *He demonstrated the effectiveness of aerial bombardment of naval warships off Hatteras, NC in September 1923.*

If I had to choose, I would rather have birds than airplanes.
> — *Charles Lindbergh*

I do not want to be a fly! I want to be a worm!
> — *Charlotte Perkins Stetson Gilman*
>> In *A Conservative*
>> *Apparently, not everyone wants to go to the stars.*

Keep'em flying.
> — *Harold N. Gilbert*
>> In a poster caption
>> *Slogan of the Air Forces during World War II.*

When the weight of the paper equals the weight of the airplane, only then you can go flying.
> — *attributed to Donald Douglas*
>> *Remarking about the amount of design and regulatory paperwork needed to make the DC-3, the first popular passenger plane.*

Speed is life; altitude is life insurance.
> — *Pilot's proverb*

The propeller is just a big fan in the front of the plane to keep the pilot cool. Want proof? Make it stop; then watch the pilot break out into a sweat.
— *Anonymous*

Astronautical Engineering

I believe the nation should commit itself to achieving the goal, before this decade is out, of landing a man on the Moon and returning him safely to Earth.
— *President John F. Kennedy*
In a special message to a joint session of Congress on May 25, 1961.

The Eagle has landed.
— *Neil Armstrong, July 21, 1969*
Announcing man's first landing on the Moon.

Here's one small step for a man, one giant leap for mankind.
— *Neil Armstrong, 1969*
Spoken while being the first person to step upon the Moon.

NASA discovers there is an edge to the universe.

From now on, we live on a world where man has walked on the moon.
— *Jim Lovell*
Astronaut Jim Lovell is referring to the 1969 Apollo 11 moon landing.

For the first time in all of time, men have seen the Earth: seen it not as continents or oceans from the little distance of a hundred miles or two or three, but seen it from the depths of space; seen it whole and round and beautiful and small... a tiny raft in the enormous empty night.
— *Archibald MacLeish*
In the **New York Times**, December 26, 1968
Referring to the photos taken by Apollo 8 on Christmas Eve which vividly illustrated the limit of the resources that we possess on his "tiny raft."

Earth is the cradle of mankind, but man cannot live in the cradle forever.
— *Konstantin E. Tsiolkovsky*
An early Russian rocket theorist.

Some say God is living there [in space]. I was looking around very attentively, but I did not see anyone there. I did not detect either angels or gods... I don't believe in God. I believe in man—his strength, his possibilities, his reason.

> *— Gherman Titov*
> Quoted in *The Seattle Daily Times*, May 6, 1962
> *Soviet Cosmonaut in comments made at the World's Fair in Seattle, Washington.*

The exploration and ultimate colonization of the solar system is the only future worthy of truly great nations at this time in history. The Soviets, who cannot even feed themselves, seem to understand this.

> *—John S. Powers*

Now I know what a billion dollars represents.

> *— Wernher von Braun*
> As told to Bill Smollen by Wernher von Braun
> *Spoken to a reporter about the tremendous national effort on behalf of the Apollo space program.*

Astronomy compels the soul to look upwards and leads us from this world to another.

> *— Plato*
> In *The Republic*, circa 390 B.C.E.

Nicolas Copernicus (1473–1543) appeared on the Polish 1000 Zloty note. He was the first modern scientist to propose a model of the solar system in which the sun was at the center. He suppressed publications of his work during his lifetime.

A physical cosmography, or picture of the universe, should begin, not with the earth, but with the regions of space—the distribution of matter in the universe.

> *— Alexander von Humboldt*
> In *Cosmos, a Sketch of the Universe*, 1848

We see matter existing in space…in the form of spheroids…and in the form of self-luminous vapour dispersed in shining nebulous spots or patches.
— *Alexander von Humboldt*
In *Cosmos, a Sketch of the Universe*, 1848
Speaking about nebulae.

Black holes are where God divided by zero.
— *Stephen Wright*

Galileo Galilei appeared on the 1976 Italian 2000 Lire note. In some sense, he was the first modern scientist. He made critical discoveries about the dynamics of moving bodies.

The eternal silence of these infinite spaces alarms me.
— *Blaise Pascal*
In *Pensees*, 1670

Space, the final frontier.
— *Gene Roddenberry*
In the introduction to *Star Trek*, 1966

Don't tell me that man doesn't belong there. Man belongs wherever he wants to go; and he'll do plenty well when he gets there.
— *Wernher von Braun*
Developer of the German rocket program which sent V-2 rockets into London during WWII. After the war, he led the American rocket program.

Fate has ordained that the men who went to the moon to explore in peace will stay to rest in peace.
> — *William Safire*
> In a unread speech written for President Richard Nixon, July 18, 1969
> *The opening line of a speech written in case the astronauts of Apollo XI were unable to return to Earth.*

It [the space program] will free man from his remaining chains, the chains of gravity which still tie him to this planet. It will open to him the gates of heaven.
> —*Wernher von Braun, February 10, 1958*

This young scholar has produced a rocket that upsets all known ballistic laws. I am convinced that this young scientist is right when he says that in his opinion more powerful rockets would be capable of exploring the space surrounding the earth and perhaps even several planets in our solar system. We will have von Braun to thank for the uncovering of a great many secrets.
> — *Adolf Hitler*
> Quoted by Otto Skorzeny in his memoirs *My Commando Operations, The Memoirs of Hitler's Most Daring Commando*
> *Description of Wernher von Braun after a discussion with von Braun regarding the V-2 rocket.*

No one regards what is before his feet; we all gaze at the stars.
> — *Quintus Ennius*
> In *Iphigenia*. Quoted by Cicero in *De Divinatione*, circa 70 B.C.E.

God has no intention of setting a limit to the efforts of man to conquer space.
> — *Pope Pius XII*

Professor Goddard does not know the relation between action and reaction and the need to have something better than a vacuum against which to react. He seems to lack the basic knowledge ladled out daily in high schools.
> — *New York Times editorial, 1921*
> *Speaking about Robert Goddard's revolutionary rocket work.*

Man will never reach the moon regardless of all future scientific advances.
> — *Lee De Forest, February 25, 1967*

The pitted appearance of the lunar surface, which one observer has likened to that of a WWI battlefield, is enhanced by the presence of long shadows.

— *Eugene M. Shoemaker, 1966*
Quoted in *Exploring Space with a Camera* by Edgar M. Cortright, 1968.
Comments about lunar photographs taken by Surveyor I in June–July, 1966.

Surveyor I stands physically on the Moon, an enduring monument to its creators, a solitary artifact of men who live on another body of the solar system, but its true resting place is in the pages of history, where even now is being inscribed man's conquest of space.

— *Homer E. Newell, 1966*
Comments about the unmanned U.S. spacecraft Surveyor I. The mission of the spacecraft was cut short on July 13, 1966 due to battery failure.

Soon now, man will tread the barren wastes of the Moon. One day, after suitable reconnaissance, Mars will yield its secrets in like manner. Space stations will perpetually circle the Earth in the conduct of both scientific observations and technological investigations. And precursor spacecraft will probe the outer regions of the solar system, and peer tentatively toward the starry, endless universe beyond.

— *Edgar M. Cortright*
In *Exploring Space with a Camera*, 1968

It is possible that radioactive material exists on the moon that has to be procured on earth at great expense… like uranium and thorium and other nuclear fission material.

— *Leslie Greener*
In *Moon Ahead*, 1951

Doesn't that idiot paper hanger understand that a rocket is like a tree— it needs some time to grow.

— *Wernher von Braun*
As told to Bill Smollen
Spoken about Adolf Hitler after Hitler made a crude comment about the early V-2 rocket. Von Braun escaped execution for this comment by joining the Nazi party.

There is just one thing I can promise you about the outer-space program: your tax dollar will go farther.

— *Wernher von Braun*

Hitch your wagon to a star.
— *Ralph Waldo Emerson*
In *Society and Solitude*, 1870

Living on Earth may be expensive, but it includes an annual free trip around the Sun.
— *Bumper sticker*

Space isn't remote at all. It's only an hour's drive away if your car could go straight upwards.
— *Fred Hoyle*

Electrical Engineering

Why sir, there is every possibility that you will soon be able to tax it!
— *Michael Faraday*
Quoted in *Democracy and Liberty* by W.E.H. Lecky, 1899
In reply to Mr. Gladstone, the future British Prime Minister, when asked about the usefulness of electricity.

Temples of kilowatts
— *Yevgeny Yevtushenko*
In *National Geographic Magazine*
Speaking about hydroelectric dams.

Nikola Tesla (1856–1943) was born in Croatia and immigrated to America. He contributed to the development of electrical technology. He is featured on the front of the 5 million Dinar Note.

The reverse features one of the hydroelectric projects which was enabled by his work.

Coal is a portable climate. It carries the heat of the tropics to Labrador and the polar circle; and it is the means of transporting itself whithersoever it is wanted. Watt and Stephenson whispered in the ear of mankind their secret, that *a half-ounce of coal will draw two tons a mile*, and coal carries coal, by rail and by boat, to make Canada as warm as Calcutta, and with its comfort brings its industrial power.

> — *Ralph Waldo Emerson*
> In *The Conduct of Life*, "Wealth," 1860

Wonderful as are the laws…of electricity when made evident to us in inorganic…matter, their interest can bear scarcely any comparison… when connected with the nervous system and with life.

> — *Michael Faraday*
> In *Experimental Researches in Electricity*, 1844-1855
> *Faraday is the founder of the modern science of electricity and discovered the relationship between electric fields and magnetic fields which forms the operating principle of the electric motor. The nervous system also runs on a form of electricity.*

Michael Faraday appears on the 1991 British 20 pound note. Faraday was one of the primary discoverers of the properties of electricity and magnetism and their relationship.

Is it a fact—or have I dreamt it—that, by means of electricity, the world of matter has become a great nerve, vibrating thousands of miles in a breathless point of time?

> — *Nathaniel Hawthorne*
> In *The House of Seven Gables*, spoken by Clifford Pyncheon, 1851

Guglielmo Marconi, the developer of the first successful radio, appears on the 1990 Italian 2000 Lire note. The reverse of the note features early radio equipment.

The new electronic interdependence recreates the world in the image of a global village.
— *Marshall McLuhan*
In the *Gutenberg Galaxy,* 1962

Energy is eternal delight.
— *William Blake*
In *The Marriage of Heaven and Hell,* 1790

You see, wire telegraph is a kind of a very, very long cat. You pull his tail in New York and his head is meowing in Los Angeles. Do you understand this? And radio operates exactly the same way: you send signals here, they receive them there. The only difference is that there is no cat.
— *attributed to Albert Einstein*
Einstein's reply when asked to describe radio.

More Power! Augh, Augh, AUUUGHHH!
— *Tim Allen*
In *Home Improvement,* 1992

Her own mother lived the latter years of her life in the horrible suspicion that electricity was dripping invisibly all over the house.
— *James Thurber*
In *My Life and Hard Times,* 1933

Atomic Power

The most vitally interesting question which the physics of the future has to face is, Is it possible for man to gain control of subatomic energy and to use it for his own ends? Such a result does not now seem likely or even possible; and yet the transformations which the study of physics has wrought in the world within a hundred years were once just as incredible as this. In view of what physics has done, is doing, and can yet do for the progress of the world, can any one be insensible either to its value or to its fascination?

> — *Robert A. Millikan and Henry G. Gale*
> In *A First Course in Physics*, 1906
> *With the discovery of natural nuclear reactions in radium in 1903, the prospect of a nuclear age was an appealing possibility.*

That is how the atom is split. But what does it mean? To us who think in terms of practical use it means—Nothing!

> — *Ritchie Calder*
> In *The Daily Heard*, June 27, 1932
> *This quote illustrates why basic research must be funded. The practical uses of fundamental discoveries are often difficult to know until decades after their discovery.*

Some recent work by E. Fermi and L. Szilard, which has been communicated to me in manuscript, leads me to expect that the element uranium may be turned into a new and important source of energy in the immediate future.

> — *Albert Einstein*
> In a letter to President Roosevelt, August 2, 1939. Quoted in *Einstein on Peace* by Otto Nathan and Heinz Norden
> *In this letter, Einstein was encouraging the development of the atomic bomb.*

The bomb will never go off, and I speak as an expert in explosives.

> — *Admiral William Daniel Leahy, 1945*
> Quoted in *Time Magazine*
> *Advice given to President Truman on the atom bomb project.*

The release of atomic energy constitutes a new force too revolutionary to consider in the framework of old ideas.

> — *President Harry S. Truman*
> In a message to Congress on atomic energy, October 3, 1945

Our children will enjoy in their homes electrical energy too cheap
to meter.
> — *Lewis Strauss*
> In a speech to the National Association of Science Writers, September 16,
> 1954. Quoted in the *New York Times*.
> *Chairman of the Atomic Energy Commission*

Breeder reactors will be the backbone of an emerging nuclear economy
and plutonium will be a logical contender to replace gold as the standard of our monetary system.
> — *Glenn Seaborg, 1970*
> Quoted in *Harper's Magazine*

The Golf Manor Superfund cleanup was provoked by the boy next door,
David Hahn, who attempted to build a nuclear breeder reactor in his
mother's potting shed as part of a Boy Scout merit-badge project.
> — *Ken Silverstein*
> In *Harpers Magazine*, November 1998
> *In 1994, seventeen-year-old David Hahn was able to create a small-scale
> breeder reactor by building a radium and americium powered neutron gun,
> a tritium-based nuclear booster, and home-refined thorium and uranium
> blanket. David received his merit-badge in Atomic Energy.*

The vast energy locked with the hearts of the atoms of matter was
released for the first time in a burst of flame such as had never before
been seen on this planet.
> — *William L. Laurence*
> In the *New York Times*, September 26, 1945
> *Description of the first atomic blast in New Mexico which was conducted
> on July 16, 1945.*

Nature is neutral. Man has wrested from nature the power to make the
world a desert or to make the deserts bloom. There is no evil in the
atom; only in men's souls.
> — *Adlai Stevenson*
> In a speech in Hartford, Connecticut, September 18, 1952

The test is pointless. It will kill people for no reason.
> — *Andrei Sakharov*
> In a letter to Soviet Prime Minister Nikita Khrushchev
> *Warning about the dangers of radiation fallout from the testing of nuclear
> weapons. Sakharov was later exiled for this and other outspoken views.*

Parker Pen Company technicians have visualized the atomic fountain pen for the year 1975. It may carry in its protective barrel a tiny pellet of an atomically active substance… A red sapphire bearing would focus the pen while it would burn a written impression on paper.
— *The American Educator Encyclopedia, 1952*

Computer Science

I do not fear computers. I fear the lack of them.
— *Isaac Asimov*

A computer on every desk and in every home.
— *Bill Gates*

It is unworthy of excellent men to lose hours, like slaves, in the labors of calculation.
— *Gottfried Wilhelm Leibnitz*

Computers make it easier to do a lot of things, but most of the things they make it easier to do don't need to be done.
— *Andy Rooney*

The Computer is incredibly fast, accurate and stupid. Man is unbelievably slow, inaccurate and brilliant. The marriage of the two is a challenge and an opportunity beyond imagination.
— *Walesh, 1989*
Summarizing the reasons for using computer modeling for hydrologic and water quality analysis.

The best computer is a man, and it's the only one that can be mass-produced by unskilled labor.
— *Wernher von Braun*
But society spends a lot on quality control!

Compassion—that's the one thing no machine ever had. Maybe it's the one thing that keeps man ahead of them.
— *D.C. Fontana*
In *Star Trek: The Ultimate Computer*, 1968

The greatest task before civilization at present is to make machines what they ought to be, the slaves, instead of the masters of men.
> — *Havelock Ellis*
> In *Little Essays on Love and Virtue*

It would be very discouraging if somewhere down the line you could ask a computer if the Riemann hypothesis is correct and it said, "Yes, it is true, but you won't be able to understand the proof."
> — *Ronald Graham*
> Quoted in *Scientific American* by John Horgan

One day ladies will take their computers for walks in the park and tell each other, "My little computer said such a funny thing this morning."
> — *Alan Turning*
> Quoted in *Time Magazine*

Programming today is a race between software engineers striving to build bigger and better idiot-proof programs, and the Universe trying to produce bigger and better idiots. So far, the Universe is winning.
> — *Rich Cook*

A good calculator does not need artificial aids.
> — *Lao Tzu*
> In *Tao Te Ching*, circa 500 B.C.E.
> *Speaking about the abacus.*

I cannot do it without counters.
> — *William Shakespeare*
> In *The Winter's Tale*, 1610

Computers are composed of nothing more than logic gates stretched out to the horizon in a vast numerical irrigation system.
> — *Stan Augarten*
> In *State of the Art: A Photographic History of the Integrated Circuit*

The successful companies of the next decade will be the ones that use digital tools to reinvent the way that they work.
> — *Bill Gates*
> In *Business @ the Speed of Thought*, 1999

The purpose of computing is insight, not numbers!
> — *R. W. Hamming*
> Quoted in "Mathematics and Modern Technology" by R.S. Pinkham in *The American Mathematical Monthly*

If the 1980s were about quality and the 1990s were about re-engineering, then the 2000s will be about velocity. About how quickly business itself will be transacted.
> — *Bill Gates*
> In *Business @ the Speed of Thought*, 1999

The iron machines still exist, but they obey the orders of weightless bits.
> — *Italo Calvino and Patrick Creagh*
> In *Six Memos for the Next Millennium/the Charles Eliot Norton Lectures 1985–86*, 1993

It would appear that we have reached the limits of what it is possible to achieve with computer technology, although one should be careful with such statements, as they tend to sound pretty silly in five years.
> — *John von Neumann, 1949*

I think there is a world market for maybe five computers.
> — *Thomas Watson, 1943*
> Chairman of IBM. Does this attitude help to explain why IBM had trouble when personal computers started to proliferate?

There is no reason anyone would want a computer in their home.
> — *Ken Olson, 1977*
> President and founder of Digital Equipment Corporation. Perhaps this helps explain why DEC had trouble during the boom in PCs and is now out of business.

640K ought to be enough for anybody.
> — *attributed to Bill Gates, 1981*
> Founder of Microsoft Corporation setting the upper limit for how much memory the DOS operating system could access.

To err is human but to really foul things up requires a computer.
> — *Anonymous*
> In the *Farmers' Almanac*, 1978

On two occasions I have been asked [by members of Parliament], "Pray, Mr. Babbage, if you put into the machine wrong figures, will the right answers come out?" I am not able rightly to apprehend the kind of confusion of ideas that could provoke such a question.
> — *Charles Babbage*
> The modern phrasing is GIGO: Garbage In, Garbage Out.

No offense to your dots and dashes, Mr. Morse, but how much can you ever do with just a binary code?
> — *Bob Thaves*
> As a cartoon caption in **Frank & Ernest**, 1999

A modern computer hovers between the obsolescent and the nonexistent.
> — *attributed to Sydney Brenner*
> In **Science**, January 5, 1990

Transportation

Let the street be as wide as the height of the houses.
> — *Leonardo da Vinci*
> In his notebooks

Our national flower the concrete cloverleaf.
> — *Lewis Mumford*

Suburbia: Where they tear out the trees and then name streets after them.
> — *Anonymous*

Our nature consists in movement; absolute rest is death.
> — *Blaise Pascal*
> In **Pensees**, 1670
> This gives new meaning to the saying that "rush hour traffic was murder."

Our road is not built to last a thousand years, yet in a sense it is. When a road is once built, it is a strange thing how it collects traffic, how every year as it goes on, more and more people are found to walk thereon, and others are raised up to repair and perpetuate it, and keep it alive.
> — *Robert Lewis Stevenson*
> In an address on the opening of the **Road of Gratitude**, October, 1894
> Many of the roads and bridges originally built by the Romans are still in use today.

There can be little doubt that in many ways the story of bridge building is the story of civilization. By it, we can readily measure the important part of a people's progress.

> — *Franklin D. Roosevelt, October, 1931*
>> *Then Governor of New York speaking about the construction of the George Washington Bridge. The main span of this bridge was about twice the length of any bridge in existence at the time.*

Historic bridges are important links to our past, serve as safe and vital transportation routes in the present, and can represent significant resources in the future.

> — *Surface Transportation and Uniform Relocation Assistance Act, 1987*
>> *This act provided funds for the preservation of historic bridges.*

On the reverse of the Brazilian 1990 Hungarian 500 Forint note which honors the Hungarian poet Ady Endre is a picture of one of the suspension bridges over the Danube.

Discriminating persons believe that such a structure would prove an eyesore to those now living and a betrayal of future generations for whom the current generation is a trustee. A bridge of the size projected—the plans call for towers 800 feet high—would certainly mar if not utterly destroy the natural charm of the harbor famed throughout the world.

> — *"The Wasp," May 2, 1925*
> Quoted in **Great American Bridges and Dams**
>> *San Francisco newspaper editorial expressing opposition to the Golden Gate Bridge in San Francisco. The bridge is now considered to be a landmark and a symbol of San Francisco.*

There is no part of the American coast where vessels are more exposed to shipwreck than when they are passing along the shores of North Carolina.

> — *A Congressional Committee, 1806*
>> *This report resulted in major federal funding for lighthouse construction.*

Little Tin Gods on Wheels.
> — *Rudyard Kipling*
> In *Public Waste*, 1886
> *An apt description of the modern automobile driver.*

The bicycle craze has hit the town, and it is fast becoming evident that the era of the horse is disappearing.
> — In the *Rockingham Southern Index*, June 15, 1886

Sometimes the road less traveled is less traveled for a reason.
> — *Jerry Seinfeld*

Civil Engineering

Our property...is so close to Babylon that we enjoy all of the advantages of the city, and yet when we come home we are away from all of the noise and dust
> — *Letter inscribed on a clay tablet to the King of Persia, 539 B.C.E.*

What a time! What a civilization!
> — *Marcus Tillius Cicero*
> In reference to the Roman Empire

Illustration of Emperor Yu surveying prior to starting river channelization.

Respect the ancient system and do not lightly modify it.
> — *Inscription on the Li Erlang Temple*, 240 B.C.E.
> Quoted in *The Builders—Masters of Engineering*
> *The inscription is referring to the Guan Xian flood control project along the Min River. It seems that ancient engineers had concerns similar to modern engineers regarding the maintenance and operation of projects.*

There is no place more delightful than home.
— *Marcus Tillius Cicero*

...a foolish man who builds his house upon the sand.
— *Matthew 7:26*
While not intended as divine geotechnical guidance, the thought is applicable to a variety of situations where the importance of a good foundation is needed.

I found Rome a city of bricks and left it a city of marble.
— *Augustus Caesar*
In *Augustus*
But how did he find Egypt, and in what state was it left?

I love the smell of fresh concrete in the morning!

The civilization that pours the most concrete has the greatest impact on future civilizations.
— *Michael Fripp, 1999*
All of the ancient cultures that strongly influenced the modern culture, such as the Romans and the Greeks, spent enormous investments building structures. The cultures that did not build structures are largely forgotten.

They left no writing, but they did leave all those houses.
— *Henry Glassie, 1975*
Quoted in *Archaeology* by David Thomas
American structural anthropologist discussing his research on colonial housing.

It represents the closest thing to the scientific mass production of houses that the United States has ever seen.
— *The American Educator Encyclopedia, 1952*
In reference to Levittown, New York, where each house was a duplicate of every other.

The military engineer had died and his close relative, the civil engineer, had taken his place.
— *Anonymous*
Quoted in *The Corps of Engineers: The War Against Germany*
An American Regimental Commander on the eve of World War II referring to the expanding duties required of the engineers in the military.

Bunch of damn good engineers.

— *Colonel James Stratton, 1942*

Referring to the design team who were instrumental in advancing pavement design and soil analysis for road and airfield construction in WWII. Much of the research done by this group is still in use today.

I know these Army Engineers very well… I venture to say that in your whole experience you have never met men of a higher standard of character, of a higher devotion to public duty, and of a greater skill in their profession.

— *President William Taft*

Quoted in **Of Men and Rivers—The Story of the Vicksburg District** *by Gary Mills*

Shall it [tax revenues] lie unproductive in the public vaults? Shall the revenue be reduced? Or shall it not rather be appropriated to the improvement of roads, canals, rivers, education, and other great foundations of prosperity and union.

— *President Thomas Jefferson*

The reverse of the Brazilian 1000 Mil Cruzeiros note features a stylistic representation of surveying equipment on top of a topographical map.

I don't give a damn whether a project is feasible or not. I'm getting the money out of Congress, and you'd damn well better spend it. And you'd better be here early tomorrow morning ready to spend it, or you may find someone else at your desk!

— *Michael W. Straus, 1952*

Quoted in **Cadillac Desert** *by Marc Reisner*

Commissioner of the Bureau of Reclamation in a 1952 'pep talk' to the Billings, Montana regional office.

Look upon my works, ye Mighty,
and despair!
> — *Percy Bysshe Shelley*
> In "Ozymandias," 1818
>> *The poem was inspired by a legend that this quote was carved on Ramses II's tomb. The poem continues by stating that nothing else remained of the tomb except the base with this quote.*

A building's natural strength should
be expressed.
> — *Faziur Khan*
> Quoted in *The Builders—
> Masters of Engineering*

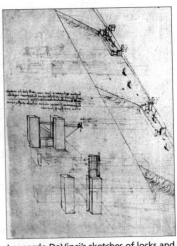

Leonardo DaVinci's sketches of locks and dams. (Codex Atlanticus, f. 90 v/33 v-a)

The larger our great cities grow, the more irresistible becomes the attraction which they exert on the children of the country, who are fascinated by them, as the birds are fascinated by the lighthouse or the moths by the candle.
> — *Havelock Ellis*
> In *The Task of Social Hygiene*

It [Architecture] is… among the most important arts; and it is desirable to introduce taste into an art which shows so much.
> — *Thomas Jefferson, 1788*

In the architectural structure, man's pride, man's triumph over gravitation, man's will to power, assume a visible form. Architecture is a sort of oratory of power by means of forms.
> — *Friedrich Wilhelm Nietzsche*
> In *The Twilight of the Idols*, 1888

The tallest structure indicates the dominant power in a community.
> — *Michael Fripp, 1999*
>> *In antiquity, trees were the tallest structure and nature was the dominant element. In time, churches dominated over the trees, then factories became taller than the churches, and now office buildings are taller than the factories.*

The Promised Land always lies on the other side of a wilderness.
> — *Havelock Ellis*
> In *The Dance of Life*

I must borrow a word from Hollywood: the job is colossal.
> — *General Eugen Reybold*
> Quoted in *Builders and Fighters*
> *Referring to the buildup of the US Army Corps of Engineers in the early days of WWII.*

I have lately been surveying the Walden woods so extensively and minutely that now I see it mapped in my mind's eye.... I fear this particular dry knowledge may affect my imagination and fancy, that it will not be easy to see so much wildness and native vigor there was formerly.
> — *Henry David Thoreau, 1858*
> *Thoreau became a surveyor to make ends meet. As he became accomplished in this craft, he began to become concerned that these new skills would be incompatible with the skills of a poet. He appears truly to be concerned that a surveyor could not "see the forest for the trees."*

In the United States there is more space where nobody is than where anybody is. This is what makes America what it is.
> — *Gertrude Stein*
> In *The Geographical History of America*, 1936

The [workmen]... are in the water constantly and require whiskey not only as a protection against sickness but to stimulate them to activity and perseverance. Will the Government allow me to get a supply to issue daily by myself?
> — *Clement Smith, 1871*
> Quoted in *Of Men and Rivers—The Story of the Vicksburg District* by Gary Mills
> *An engineer on the first major government-sponsored clearing and snagging operation on the Ouachita and Little Missouri.*

Three engineering students were gathered together discussing the possible designers of the human body.

One said, "It was a mechanical engineer. Just look at all the joints."

Another said, "No, it was obviously an electrical engineer. The nervous system has many thousands of electrical connections."

The last said, "Actually, I think it was a civil engineer. Who else would run a toxic waste pipeline through a recreational area?"
> — *Anonymous*

Hydraulic Engineering

Big whorls have little whorls which feed on their velocity
And little whorls have lesser whorls and so on to viscosity.
> — *Lewis F. Richardson*
> *1920s English scientist who studied fluid turbulence by throwing sacks of white parsnips into the Cape Cod Canal*

When you try to explain the behavior of water, remember to demonstrate the experiment first and the cause next.
> — *Leonardo da Vinci*

I really think He may have
an answer to the first question.
> — *Werner Heisenberg*
> *On his deathbed, the quantum theorist Werner Heisenberg declared that he would have two questions for God: why relativity and why turbulence.*

From a drop of water a logician
could predict an Atlantic or a Niagara.
> — *Sir Arthur Conan Doyle*
> In *A Study in Scarlet*, 1929

Inequality is the cause of all
local movements.

Study of whirlpools by Leonardo da Vinci. (RL 12660v)

> — *Leonardo da Vinci*
> *Actually, the inequality of forces causes accelerations, not movements.*

Where the flow carries a large quantity of water, the speed of the flow is greater and vice versa.
> — *Leonardo da Vinci*
> *This is the statement of continuity.*

From the twin viewpoints of quantity and quality, water-resources projects are of paramount importance to the maintenance and progress of civilization as it is known today.
> — *Richard French*
> In *Open Channel Hydraulics*, 1985

The remarkable levels of public health and safety enjoyed by the urban population of the developed world are due in considerable part to investments in hydrology over the past century.
> — *R. L. Bras and P. S. Eagleson*
> In an editorial in *EOS, Transactions, American Geophysical Union* titled "Hydrology, the Forgotten Earth Science"

It may be permitted to doubt whether the public at large has any sense of the part which fluid mechanics plays, not only in our daily lives, but also throughout the entire domain of Nature. Matter as we know it is either solid or fluid...Thus the flow of rivers and streams in their boundaries; the circulation of the blood in our arteries and veins; the flight of the insect, the bird, and the airplane; the movement of a ship in the water or of a fish in the depths. These are all, in major degree, varied expressions of the laws of fluid dynamics.
> — *W. F. Durand*
> In "The Outlook of Fluid Mechanics" in the *Journal of the Franklin Institute*, August, 1939

There are... four [things] which I know not: The way of an eagle in the air; the way of a serpent upon the rock; the way of a ship in the midst of the sea; and the way of a man with a maid.
> — *Proverbs 30:18-19*

Mere toys for youngsters of the profession.
> — *Anonymous*
> Quoted in *Builders and Fighters*
> *Derisive reference used in 1930 to the idea of using physical models in river engineering.*

A solid heavier than a fluid will, if placed in it, descend to the bottom of the fluid, and the solid will, when placed in the fluid, be lighter than its true weight by the weight of the fluid displaced.
> — *Archimedes*
> Quoted in *The Words of Archimedes* by T.L. Heath

The resistance arising from the want of lubricity in the parts of a fluid, is, other things being equal, proportional to the velocity with which the parts of the fluid are separated from one another.
> — *Sir Isaac Newton*
> *Newton's model of viscosity.*

The practical, full-scale aspect of the problems are of greater importance and are more dependent upon varying and uncertain field conditions than is common in other branches of engineering. Field experience in the solution of problems of this nature is undoubtedly of much greater value than laboratory experiments could possibly be, and the application of principles in the laboratory to the solution of practical problems in the field must be difficult and uncertain.

> — Letter prepared for Secretary of War Dwight W. Davis, 1926
> *The letter expressed opposition to the funding of the Army Corps of Engineers Waterways Experiment Station. The applications of the fundamental principle of non-dimensional scaling were not brought into widespread practical use until the 1940s.*

In the interests of the national welfare there must be national control of all running waters of the United States, from the desert trickle that might make an acre or two productive to the rushing flood waters of the Mississippi.

> — *National Resources Planning Board, 1934*

The fundamental principle of river engineering is to arrange the works so that the current is utilized to the greatest extent possible to move the bed material and to produce the desired channel conditions.

> — *Armin Shoklitsch*
> In *Hydraulic Structures–A Text and Handbook*, 1937

The alignment of a proposed river channel must be chosen especially carefully because it has a decisive effect on the configuration of the bed.

> — *Armin Shoklitsch*
> In *Hydraulic Structures–A Text and Handbook*, 1937
> *We now know that a channelized stream often reverts to its natural serpentine pattern.*

Streamflow quantity and timing are critical components of water supply, water quality, and the ecological integrity of river systems.

> — *N.L. Poff, J.D. Allen, M.B. Bain, J.R. Karr, K.L. Prestegaard, B.D. Richter, R.E. Sparks, and J.C. Stromberg*
> In "The Natural Flow Regime" in *Bioscience*, 1997

The military engineers are taking upon their shoulders the job of making the Mississippi River over again, a job transcended in size only by the original job of creating it.

> — *Mark Twain*

Consider now the work of God, for who can make that straight which He hath made crooked.

> — *Ecclesiastes 7:13*
>
> *While perhaps not intended as a summary of river mechanics, the quote is an apt assessment of river behavior. A river often meanders and becomes more crooked until the river breaks through the crook, straightens the stream, and forms an ox-bow from the crook. Even a river that has been artificially straightened may form meanders.*

Irrigation

May your canal be filled with sand.

> — *Ancient Babylonian Curse*
> Quoted in *The Ancient Engineers* by L. Sprague De Camp
>
> *This quote shows us two things. First, curses were much more interesting a few millennia ago. Second, excessive aggradations have long been realized as a problem for constructed waterways.*

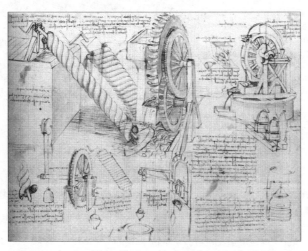

Drawings of water lifting devices. (Codex Atlanticus, f. 26 v/7 v-a)

Let not a single drop of water that falls on the land go into the sea without serving the people.

— *Parakrama Bahu I, circa 1100*
Quoted by Al Gore in *Earth in the Balance—Ecology and the Human Spirit*

Irrigation of the land with seawater desalinated by fusion power is ancient. It's called "rain."

— *Michael McClary*

Today we play a brand new ball game. Water, in the right place, at the right time, sufficient and adequate in quality, is taken for granted by millions of Americans—until the supply fails.

— *Colonel John Brennan*
Quoted in *Of Men and Rivers—The Story of the Vicksburg District* by Gary Mills

Of all the works of civilization that interfere with the natural water distribution system, irrigation has been by far the most pervasive and powerful.

— *Al Gore*
In *Earth in the Balance—Ecology and the Human Spirit*, 1992

Illustration of the construction of Emperor Yu's irrigation canal (2000 B.C.E.) through Mount Longmen.

Ours was the first and will doubtless be the last party of whites to visit this profitless locale.

— *Lt. Joseph Christmas Ives, 1857*
Quoted in *Cadillac Desert* by Marc Reisner
On sailing up the Colorado River to a point near the present location of Las Vegas.

Wherever the hand of Los Angeles has touched Owens Valley, it has turned back into desert.

— *Fred Eaton, 1924*
Quoted in *Cadillac Desert* by Marc Reisner
Former mayor of Los Angeles speaking about the effects of the various projects used to provide water for the city.

The water necessary for greatly expanded irrigation development will
be provided, at whatever cost may be required.

> — *A. D. Edmonston, 1951*
> Quoted in *Cadillac Desert* by Marc Reisner
> *In an inventory of California's water resources.*

In order to maintain and ensure the long-term viability of irrigated
agriculture and to provide enough water to carry the salts to the ocean
or some other natural sink, the development of water resources should
be intensified.

> — *Arthur Pillsbury, 1951*
> Quoted in *Cadillac Desert* by Marc Reisner
> *In a* Scientific American *article warning of the accumulation of salts in
> irrigated fields.*

The United States has virtually set up an empire on impounded and
redistributed water. The nation is encouraging development, on a scale
never before attempted, of lands that are almost worthless except for
the waters that can be delivered to them by the works of man.

> — *Charles P. Stevens, 1946*
> Quoted in *Cadillac Desert* by Marc Reisner

It's but little good you'll do a-watering the last year's crop.

> — *George Eliot*
> In *Adam Bede*, 1859

Flood Control

We, the Nation, must build the levees and build them better and more
scientifically than ever before.

> — *President Theodore Roosevelt*
> Quoted in *Of Men and Rivers—The Story of the Vicksburg District*
> by Gary Mills

We set out to tame the rivers and ended up killing them.

> — *Marc Reisner*
> In *Cadillac Desert*, 1986

Clear out the beds and keep the dykes and spillways low.

— *Li Bings, 240 B.C.E.*
Quoted in *The Builders—Masters of Engineering*

*Instructions for the maintenance of the Guan Xian flood control project
along the Min River. This project is still in operation and the instructions are
similar to those used to upkeep modern flood control projects.*

Illustration of Emperor Yu's philosophy of flood control through
channelization of the river beds rather than restricting the
flow through the use of dikes.

The wisdom of Congress should be invoked to devise some plan by
which that great river shall cease to be a terror to those who dwell
upon its banks.

— *James Garfield, 1880*
Quoted in *Of Men and Rivers—The Story of the Vicksburg District*
by Gary Mills

*Referring to the Mississippi River in his speech accepting nomination for
Presidency of the United States.*

It is hereby recognized that destructive floods upon the rivers of the
United States,... causing loss of life and property,... constitute a menace
to the national welfare.

— *Section 1 of the Flood Control Act of 1936*

*This act is credited with being the impetus for modern federal flood control
in the United States. From 1938 to 1988, the federal government through
the U.S. Army Corps of Engineers has invested $23 billion in flood control
projects. These projects are estimated to have prevented $150 billion in
damages. Still, approximately $4 billion in flood damages and 185 deaths
occur annually.*

Practical dam safety management is intrinsically risk management.

— *Davis Bowles, 1996*

The purposes served by river works are: protection of river banks and valley lands against devastation by floods; the control of detritus and ice; the lowering or raising of the stream bed; and, in many cases, the lowering of the groundwater table. The reclamation of the valley lands in the wild stretches is frequently also accomplished at the same time.

> — *Armin Shoklitsch*
> In *Hydraulic Structures—A Text and Handbook*, 1937

Risk cannot be eliminated; therefore it must be managed.

> — *Institution of Civil Engineers*
> In their report *Whither Civil Engineering*, 1996

The accomplishment of such a [flood control] plan would, in many localities, reduce flood losses much more than the cost of construction and operation of such flood protection measures.

> — *The Floods of March 1936 in Pennsylvania*
> Prepared by the Commonwealth of Pennsylvania Department of Forests and Waters in cooperation with the United States Geological Survey, 1936.

All authorized flood control projects are directly connected with the national economy and are therefore either directly or indirectly related to the war effort, especially when it is remembered that one major flood in a large river basin, such as the Ohio or Mississippi, may easily accomplish in a few weeks at least the same amount of damage that can be caused by intensive air raids.

> — *1942 Annual Report of the Chief of Engineers*

It is an article of faith in the South that you send a politician to Washington to bring home a dam.

> — *Marc Reisner*
> In *Cadillac Desert*, 1986

A perfect symbol of man and nature in harmony.

> — *Carl Conduit*
> Quoted in *Great American Bridges and Dams* by D.C. Jackson
> *In reference to the Fontana Dam, one of the tallest dams in the U.S.. While dams provide important flood control and hydroelectric power benefits, dams also have significant environmental consequences.*

The huge dam over the Connecticut River at Hadley Falls, Mass., was completed on the 16th of last month, and the day of its completion was the day of its doom. From the first, imperfections were discovered in the work, and a breach, small at first, widened with great rapidity, until about three-fourths of the embankments burst away before the mighty mass of angry waters. The dam was constructed of immense timbers, fastened to the rocky bed of the river with iron bolts. Fault must be attributed to the principle of its construction.

   — *In* Scientific American, *December 1848*

The dam is becoming dangerous and may possibly go.

   — *The last telegraph message sent to the city of Johnstown on 31 May 1889 before the South Fork Dam failed.*
   The resulting flood destroyed much of Johnstown and killed over 2,200 people.

Q: What did the fish say when he hit a concrete wall?
A: Dam!

   — *Anonymous*

Life of the Engineer and Scientist

Ask her to wait a moment—I am almost done.

   — *Karl Friedrich Gauss*
   While working, when informed that his wife is dying.

To explain all nature is too difficult a task for any one man or even for any one age. 'Tis much better to do a little with certainty, and leave the rest for others that come after you, than to explain all things.

   — *Isaac Newton*

Engineers are more clever than artillerymen.

   — *Napoleon Bonaparte*
   Spoken to Gaspard Gourgaud at St. Helena, November 7, 1817

Those damned engineers.
> — *Lt. Colonel Joachim Peiper, 1944*
> German commander commenting upon an American engineer battalion who delayed the German advance during the Battle of the Bulge through the innovative use of their engineering skills.

We were the men who could do it, because, by God, we were getting it done.
> — *General Eugen Reybold, 1943*
> Quoted in *Builders and Fighters*
> Chief of Engineers referring to the U.S. Army Corp of Engineers transportation and support mission in World War II.

Intellectual "work" is misnamed; it is a pleasure, a dissipation, and is its own highest reward.
> — *Mark Twain*
> In *A Connecticut Yankee in King Arthur's Court*, 1889

Results! Why, man, I have gotten a lot of results. I know several thousand things that won't work.
> — *Thomas Edison*

O Diamond! Diamond! Thou little knowest the mischief done!
> — *Sir Isaac Newton*
> Quoted in *Familiar Quotations* by John Bartlett
> Said to a pet dog that knocked over a candle and set fire to some papers, representing several years' work.

It is common opinion among us in regard to beauty and wisdom that there is an honorable and a shameful way of bestowing them. For to offer one's beauty for money to all comers is called prostitution... So is it with wisdom. Those who offer it to all comers for money are known as sophists, prostitutes of wisdom.
> — *Socrates*
> Quoted in *Memorabilia* by Xenophen which was quoted in *The Doubter's Companion* by John Ralston Saul
> Description of a consultant.

It is of no small benefit on finding oneself in bed in the dark to go over again in the imagination the main outlines of the forms previously studied.
— *Leonardo da Vinci*
In his notebooks

There is one quality that characterizes all of us who deal with the sciences of the earth and its life—we are never bored.
— *Rachel Carson, 1963*
Quoted in *The House of Life* by Paul Brooks

The true men of action in our time, those who transform the world, are not the politicians and statesmen, but the scientists. Unfortunately, poetry cannot celebrate them, because their deeds are concerned with things, not persons and are, therefore, speechless.
— *W.H.Auden*
In *The Dyer's Hand*

I've handled all kinds of requisitions for you men out there, and I have never complained... but when you send me an order like the one I have here, I have to object. I'm not going to have the Comptroller General asking me why I have submitted a requisition for half a dozen breast pumps.
— *R.N.Duffey, 1938*
The Chief Clerk in response to a request for pumps that were to be used to draw water into manometer tubes as part of experiments at the Waterways Experiment Station. He eventually relented and made the request.

Is there a woman, whose form is more dazzling, more splendid than the two locomotives that pass over the Northern Railroad lines?
— *Joris Karl Huysmans*
In *Against the Grain*, 1884
With attitudes such as this, is there any wonder why engineers have trouble dating?

But the fact that some geniuses were laughed at does not imply that all who are laughed at are geniuses. They laughed at Columbus, they laughed at Fulton, they laughed at the Wright brothers. But they also laughed at Bozo the Clown.
— *Carl Sagan*

In the early years of this century, Charles Proteus Steinmetz, the great electrical engineer, was brought to General Electric's facilities in Schenectady, New York. GE had encountered a performance problem in one of their huge electrical generators.

Steinmetz was brought in as a consultant and found the problem very difficult to diagnose, but for some days he closeted himself with the generator, its engineering drawings, paper and pencil. After he departed, GE's engineers found a large "X" marked with chalk on the side of the generator casing. There also was a note instructing them to cut the casing open at that location and remove so many turns of wire from the stator. The generator would then function properly. And indeed it did.

Steinmetz was asked what his fee would be. He replied with the absolutely unheard of answer that his fee was $1000. Stunned, the GE bureaucracy then required him to submit a formally itemized invoice. They soon received it. It included two items:

Marking chalk "X" on side of generator: $1

Knowing where to mark chalk "X": $999

> — *Charles M. Vest*
> Spoken at commencement ceremonies at the Massachusetts Institute of Technology, June 4, 1999

I've spent my whole life tryin' to figure out crazy ways of doin' things. I'm tellin' ya—as one engineer to another—I can do this!

> — *Ronald D. Moore*
> In *Star Trek, The Next Generation: Relics*, 1993
> *Chief Engineer Montgomery Scott speaking to another Chief Engineer, Geordi LaForge.*

Venus was a statue made entirely of stone.
Without a stitch upon her, she was naked as a bone.
On seeing that she had no clothes, an engineer discoursed:
Why the damn thing's only concrete, and should be reinforced!

> — *The Engineer's Drinking Song*
> Quoted in *How to Get Around MIT*

A man was crossing a road one day when a frog called out to him and said: "If you kiss me, I'll turn into a beautiful princess." He bent over, picked up the frog and put it in his pocket. The frog spoke up again and said: "If you kiss me and turn me back into a beautiful princess, I will tell everyone how smart and brave you are and how you are my hero." The man took the frog out of his pocket, smiled at it, and returned it to his pocket.

The frog spoke up again and said, "If you kiss me and turn me back into a beautiful princess, I will be your loving companion for an entire week."

The man took the frog out of his pocket, smiled at it, and returned it to his pocket.

The frog then cried out, "If you kiss me and turn me back into a princess, I'll stay with you for a year and do ANYTHING you want."

Again, the man took the frog out, smiled at it, and put it back into his pocket.

Finally, the frog asked, "What is the matter? I've told you I'm a beautiful princess and that I'll stay with you for a year and do anything you want. Why won't you kiss me?"

The man said, "Look, I'm an engineer. I don't have time for a girlfriend, but a talking frog is cool."

— *Anonymous*

Technological Development

There is nothing permanent except change.
— *Heraclitus, 513 B.C.E.*

The science of today is the technology of tomorrow.
—*Edward Teller*

No idea is so antiquated that it was not once modern; no idea is so modern that it will not someday be antiquated.
— *Ellen Glasgow*

Every advance in [technology] has been denounced as unnatural while it was recent.
—*Bertrand Russell*

A new scientific truth does not triumph by convincing its opponents and making them see the light, but rather because its opponents eventually die, and a new generation grows up that is familiar with it.
— *Max Planck*
Quoted in *Husband Coached Childbirth* by Robert Bradley

We have actually in the United States attained within a hundred years, and primarily because of science and its applications, a higher standard of living for the common man than has existed in any time or place in history.
— *Robert A. Millikan*
In *The Autobiography of Robert A. Millikan*, 1950

The simplest schoolboy is now familiar with facts for which Archimedes would have sacrificed his life.
— *Ernest Renan*
In *Souvenirs D'enfance et de Jeunesse*, 1887

All the significant breakthroughs were breaks with old ways of thinking.
— *Thomas Kuhn*
In *The Structure of Scientific Revolutions*, 1962

The main stumbling block in the way of progress is and has always been unimpeachable tradition.
— *Chien-Shiung Wu*
Quoted in *Nobel Prize Women in Science* by Sharon McGrayne

The difficulty lies not in new ideas, but in escaping from old ones.
— *John Keynes*
In *The General Theory of Employment, Interest, and Money*, 1936

The desire to understand the world and the desire to reform it are the two great engines of progress.
— *Bertrand Russell*
In *Manners and Morals*, 1929

Progress imposes not only new possibilities for the future but new restrictions.
— *Norbert Wiener*
In *The Human Use of Human Beings*

The human future depends on our ability to combine the knowledge of science with the wisdom of wildness.
— *Charles Lindbergh*

Technology made large populations possible; large populations now make technology indispensable.
— *Joseph Wood Krutch*

Technology...the knack of so arranging the world that we need not experience it.
— *Max Frisch*
In *Homo Faber*, 1957

Certitude is not the test of certainty. We have been cocksure of many things that are not so.
— *Oliver Wendell Holmes*

Inventions have long since reached their limit, and I see no hope for further development.
— *Sextus Julius Frontinus, 98 C.E.*

Everything that can be invented has been invented.
— *Charles H. Duell, 1899*
Commissioner, U.S. Office of Patents

The Americans have need of the telephone, but we do not. We have plenty of messenger boys.
— *Sir William Preece, 1876*
Chief engineer of the British Post Office

Well-informed people know it's impossible to transmit the voice over wires and that were it possible to do so, the thing would be of no practical value.
— *Boston* Post, *1876*
Quote from an article concerning the arrest of a man who had been attempting to raise funds for work on a telephone.

Fundamental progress has to do with the reinterpretation of basic ideas.
— *Alfred North Whitehead*

Progress has not followed a straight, ascending line, but a spiral with rhythms of progress and retrogression, of evolution and dissolution.
— *Johann Goethe*

The idea, fantasy, or fairy tale invariably comes first. Following this is the stage of scientific investigation. Last comes the crowning achievement of the idea.
— *Konstantin Tsiolkovsky*

Down through the course of history, the mastery of a new environment, or of a major new technology, or of the combination of the two as we see in space, has had profound effects on the future of nations; on their relative strength and security; on the relations with one another; on their internal economic, social, and political affairs; and on the concepts of reality held by their people.
— In *NASA Report to the Committee on Science and Astronautics of the U.S. House of Representatives*, 1965
Description of the opportunities opening up in space.

It is only when science asks why, instead of simply describing how, that it becomes more than technology. When it asks why, it discovers Relativity. When it only shows how, it invents the atomic bomb.
— *Ursula K. Le Guin*
In "The Stalin in the Soul" in *Language of the Night*, 1979

The coming of the railroad... was not universally regarded as a good thing, and some people thought it would ruin the town and were strongly opposed to it.
— *W. N. Byers*
In the Denver newspaper the *Republican*, 1899
American pioneer referring to the 1870 completion of the railroad line in Denver.

Chapter 4

Man and the
Environment

We do not inherit the earth from our ancestors, we borrow it from
our children.
> — *Haida Indian saying*

Fill the earth and master it; and rule over the fish of the sea, the birds of
the sky, and all the living things that creep upon the earth.
> — *Genesis 1:28*
>> *Being master does not mean having license to exploit the Earth, rather it means that humanity holds stewardship of the land and is responsible for the Earth's environmental health.*

One can do noble acts without ruling earth and sea.
> — *Aristotle*

Habitat Destruction

The cutting of primeval forest and other disasters, fueled by the
demands of growing human populations, are the overriding threat to
biological diversity everywhere.
> — *Edward O. Wilson*
> In *The Diversity of Life*, 1992

More than 90 percent
of the forests of
western Ecuador
have been
destroyed
during the past
four decades. The loss
is estimated to have
extinguished or doomed
over half of the species
of the area's plants and animals. Many other biologically
diverse areas of the world are under similar assault.

The natural course of evolution

> — *Edward O. Wilson*
> In *The Diversity of Life*, 1992

We must ask ourselves if this is really what we want
to do to God's creation. To drive it to extinction?
Because extinction really is irreversible…
We can't bring them back!
> — *Stuart Prinam*
> In *National Geographic*, 1999

I find to my personal horror that I have not
been immune to naivete about exponential
functions … While I have been aware that the
interlinked problems of loss of biological diversity, tropical deforestation,
forest die back in the Northern Hemisphere and climate change are
growing exponentially, it is only this very year that I think I have truly
internalized how rapid their accelerating threat really is.
> — *Thomas E. Lovejoy, 1988*

Before humans invented agriculture there were 6 billion hectares of
forest on Earth. Now there are 4 billion, only 1.5 billion of which are
undisturbed primary forest. Half of that forest loss has occurred
between 1950 and 1990.
> — *D.H. Meadows, D.L. Meadows, and J. Randers*
> In *Beyond the Limits*, 1992

To waste, to destroy, our natural resources, to skin and exhaust the land
instead of using it so as to increase its usefulness, will result in under-
mining in the days of our children the very prosperity which we ought
by right to hand down to them amplified and developed.
> — *President Theodore Roosevelt*
> In a message to Congress, December 3, 1907

A choice to "do nothing" in response to the mounting evidence is
actually a choice to continue and even accelerate the reckless
environmental destruction that is creating the catastrophe at hand.
> — *Al Gore*
> In *Earth in the Balance—Ecology and the Human Spirit*, 1992

All of a sudden, they're playing leapfrog with a bulldozer.
> — *Al Gore*
> Quoted in *Time Magazine*, 1999
> *Referring to the continuing cycle of urban sprawl.*

Americans are finally realizing that once you lose land, you can't get it back.

— *Christine Todd Whitman*
Quoted in *Time Magazine*, 1999
New Jersey Governor referring to a state referendum that will buy a significant portion of the undeveloped land in her state.

The Walrus and the Carpenter
Were walking close at hand:
They wept like anything to see
Such quantities of sand:
"If this were only cleared away,"
They said, "it would be grand!"

— *Lewis Carroll*
In *Alice Through the Looking Glass*, 1872
While it might be grand for the Walrus and the Carpenter, it would devastate the local ecosystem.

Man's Impact on the Environment

One tragic example of the loss of forests and then water is found in Ethiopia. The amount of its forested land has decreased from 40 to 1 percent in the last four decades. Concurrently, the amount of rainfall has declined to the point where the country is rapidly becoming a wasteland.

— *Al Gore*
In *Earth in the Balance—Ecology and the Human Spirit*, 1992

It's a morbid observation, but if everyone on earth just stopped breathing for an hour, the greenhouse effect would no longer be a problem.

— *Jerry Adler*
Such a solution would also solve other problems as well!

The economic and technological triumphs of the past few years have not solved as many problems as we thought they would, and, in fact, have brought us new problems we did not foresee.

— *Henry Ford*
Founder of the Ford Motor Company on the dangers of pollution.

What is game to the rifle is game to the camera.

— *George Shiras*
Quoted in *National Geographic*
Shiras was a pioneer in the day and night photography of wildlife in the late 1800s.

We've lost our connection to the land and the outside world.

— *Jerry DeBin*
In *Time Magazine*, 1998
Alabama's coordinator of conservation education in support of hunting.

The sun and the moon and the stars would have disappeared long ago—as even their infinitely more numerous analogues on the earth beneath are likely to disappear—had they happened to be within the reach of predatory human hands.

— *Havelock Ellis*
In *The Dance of Life*

Many of the problems the world faces today are the eventual result of short-term measures taken last century.

— *Jay W. Forrester*

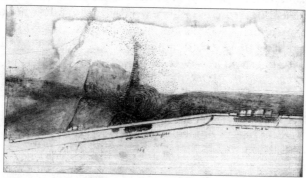

Study of erosion caused by weirs on the Arno commissioned by the Florentine government. The goal of the weirs was to gather the waters of the Arno into a single deep channel, but substantial bank erosion resulted. This is Leonardo da Vinci's sketch of the problem. (RL 12680)

All good men are for flood control and against sin. But how to control floods and what is sin—aye, there's the rub.

— *Anonymous*
Quoted in *Of Men and Rivers—The Story of the Vicksburg District* by Gary Mills
Critique of the Federal efforts in flood control at the turn of the century.

Things which matter most must never be at the mercy of things that matter least.
—*Johann Goethe*

This mess is so big and so tall there is no way we can clean it up, just no way at all.
— *Theodore Geisel, aka Dr. Seuss*
In *The Cat in the Hat*, 1957

And I brought you into a plentiful country to eat the fruit thereof and the goodness thereof; but when you entered you defiled my land, and made my heritage an abomination.
—*Jeremiah 2:17*

Man is a singular creature. He has a set of gifts which make him unique among the animals: so that, unlike them, he is not a figure in the landscape—he is a shaper of the landscape.
—*Jacob Bronowski*
In *The Ascent of Man*, 1973

From prehistory to the present time, the mindless horsemen of the environmental apocalypse have been overkill, habitat destruction, introduction of animals such as rats and goats, and diseases carried by these exotic animals.
— *Edward O. Wilson*
In *The Diversity of Life*, 1992

Human demographic success has brought the world to this crisis of biodiversity.
— *Edward O. Wilson*
In *The Diversity of Life*, 1992

The Dodo never had a chance. He seems to have been invented for the sole purpose of becoming extinct and that was all he was good for.
—Will Cuppy
In *How to Become Extinct*, 1944.
Picture from *Wood's Illustrated Natural History* by J.G. Wood, 1897.

We find ourselves, one way or another, in the midst of a large-scale experiment to change the chemical construction of the stratosphere, even though we have no clear idea what the biological or meteorological consequences may be.
— *F. Sherwood Rowland 1986*
Rowland, an atmospheric chemist, was one of the discoverers of the mechanism by which the ozone layer is eroded.

Wait a thousand years and even the garbage left behind by a vanished civilization becomes precious to us.
> — *Isaac Asimov*
> *Recycling programs seek to make use of the garbage in a far shorter time period.*

And he said,"What hast thou done? The voice of thou brother's blood crieth unto me from the ground."
> — *Genesis 4:10*

What glorious sunsets have their birth
  In Cities fouled by smoke!
This tree—whose roots are in a drain —
  Becomes the greenest Oak!
> — *William Henry Davies*
> In *Love's Rivals*
>
> *It is true; the small particulates in pollution tends to accentuate sunsets (and asthma) and the human waste in drains tends to promote the growth of plants (and cholera).*

Travels on the Water

Now for another fluvial walk.
> — *Henry David Thoreau*
> In a journal entry, July 12, 1850
> *Referring to a canoe trip.*

The most advanced nations are always those who navigate the most.
> — *Ralph Waldo Emerson*
> In *Society and Solitude*, 1870

Everyone thinks I'm a maniac. But I love this. If there's a breeze out here and there's oysters, there's nothing better.... I don't particularly care for oysters. It's the sailing I like.
> — *Jim McGlincy*
> Quoted in the newspaper the *Baltimore Sun*, January 31, 1999
> *Restorer, owner, and operator of the* Kathryn M. Lee, *the last two-masted schooner in the Chesapeake Bay oyster dredging fleet.*

If there is one kind of work which I detest more than another, it is tramping [walking]. In the North woods… you do all your sporting from your boat… This takes from recreation every trace of toil.

— *William Murray, 1869*
Quoted in "Adventures in the Wilderness; Camp Life in the Adirondacks" in *National Geographic*

The ship, a fragment detached from the earth, went on lonely and swift like a small planet.

— *Joseph Conrad*
In *The Nigger of the Narcissus*, 1898

A cup of hot cocoa would be nice...

Then the ship perished, and of them that were in it not one survived. And I was cast on to an island by a wave of the sea

— The Story of the Shipwrecked Sailor, *1700 B.C.E.*
Quoted in *The Literature of the Ancient Egyptians* by Adolf Erman.

The men could only look at each other through the falling snow, from land to sea, from sea to land, and realize how unimportant they all were.

— *Anonymous, 1839*
From a ship on the rocks in Newburyport, Massachusetts. There were no survivors.

Pleasant it is, when winds disturb the surface of the vast sea, to watch from land another's mighty struggle.

— *Lucretius*

A calm, smooth stream is a horror we all detest now.

— *O. G. Howard, 1869*
In a journal entry of the Powell Geographic Expedition. Quoted in *Cadillac Desert* by Marc Reisner
This group explored the Green River, Wyoming, in wooden dories. Several of the expedition were killed in the rapids and Howard nearly lost his life in Disaster Falls.

A good seaman can be recognized when the storm comes.

— *Greek proverb*

You are uneasy; you never sailed with *me* before, I see.
> —*Andrew Jackson*
> Remark made to an elderly gentleman who was sailing with Jackson down
> the Chesapeake Bay in an old steamboat.

There is nothing—absolutely nothing—half so much worth doing as
simply messing about in boats...or with boats...In or out of 'em, it
doesn't matter.
> —*Kenneth Grahame*
> In *The Wind in the Willows*, 1908

Voyager upon life's sea; To yourself be true, and what'er your lot may be,
Paddle your own canoe.
> —*Dr. Edward P. Philpots*
> In *Harper's Monthly*, May 1854

I fear thee, ancient mariner.
> —*Samuel Taylor Coleridge*
> In *The Rime of the Ancient Mariner*, 1798

He that's carried down the stream need not row.
> —*American proverb*

Water Quality

I desire that nobody shall conduct away any excess water with out
having received my permission...for it is necessary that a part of the
supply flowing from the water castles shall be utilized not only for
cleaning our city but also for flushing the sewers.
> —*Sextus Julius Frontinus, 98 C.E.*
> Water Commissioner of Rome

It is good to keep water in copper vessels, to expose it to sunlight, and to
filter it through charcoal.
> —*Sanskrit manuscript, 2000 B.C.E.*

Water is the most critical resource issue of our lifetime and our
children's lifetime. The health of our waters is the principal measure of
how we live on the land.
> —*Luna Leopold*

Water is one of your most urgent needs...You can't survive long without it.
> — *U.S.Army Survival Manual, FM 21-76, 1998*

Don't spit in the well for later you might wish to drink.
> — *Russian proverb*
> Quoted in *Sovietski Collection*

A well without a bucket is no good.
> — *American proverb*

Have you considered the water which you drink?
> — *The Koran 56:68*

Good to the last drop.
> — *Slogan for Maxwell House Coffee*
> Said to Joel Cheek in 1907 about Maxwell House coffee and subsequently used as an advertising slogan.

When the ditch and pond offend the nose, then look for rain and stormy blows.
> — *Old weather proverb*
> Quoted in *Air Pollution Control—A Design Approach* by C. David Cooper and F. C. Alley

The solution is dilution.
> — *Anonymous*
> This principle is also applicable to fluid mechanics and to chemistry.

Water, water everywhere, and how the boards did shrink. Water, water everywhere, Nor any drop to drink.
> — *Samuel Taylor Coleridge*
> In *The Rime of the Ancient Mariner*, 1798

Water which is too pure has no fish.
> — *Ts'ai Ken T'an*

As he brews, so shall he drink.
> — *American proverb*
> In reference to beer.

There's always good fishing in muddy waters.
— *American proverb*

The best fishing is in the deepest water.
— *American proverb.*

The stream is always purer at its source.
— *American proverb*

Power of Water

Water: In the world there is nothing more submissive and weak than water. Yet for attacking that which is hard and strong nothing can surpass it.
— *Lao-Tzu*

As a requirement for life as we know it, water has been a source of continuous preoccupation for humans since the beginning of mankind.
— *Rafael Bras*
In *Hydrology, an Introduction to Hydrologic Science*

When the well's dry, we know the worth of water.
— *Ben Franklin*
In *Poor Richard's Almanac*

Water washes everything.
— *Portuguese proverb*
Quoted in *Air Pollution Control—A Design Approach*
by C. David Cooper and F. C. Alley

Water dropping day by day will wear the hardest rock away.
— *American proverb*

Flood-protection methods are of a purely defensive nature and do nothing to mitigate the evil itself; namely the floods. Often, in fact, they make the conditions downstream worse.
— *Armin Shoklitsch*
In *Hydraulic Structures—A Text and Handbook*, 1937

Many drops of water will sink a ship.
— *American proverb*

Without water, we're all just three or four pounds of chemicals.
— *Gene Roddenberry*
In *Star Trek: The Omega Glory*, 1968

Water is the most abundant substance on earth, the principle constituent of all living things, and a major force constantly shaping the surface of the earth. It is also a key factor in air-conditioning the earth for human existence and in influencing the progress of civilization.
— *Ven Te Chow, David Maidment, and Larry Mays*
In *Applied Hydrology*, 1988

Conservation

Conservation is a state of harmony between men and land.
— *Aldo Leopold*
In *A Sand County Almanac*, 1949
In his landmark book, Leopold articulated a "land ethic" where the role of man is as a citizen of the land rather than as a conqueror of it.

We have a moral obligation to other species. The only reason for saving them is that it's right.
— *Russell Mittermeir*
Quoted in *Time Magazine*, 1998

Failure to identify and preserve endangered behavior will undermine species' potential for not only survival but evolutionary change, and hence, long-term persistence.
— *J. R. Clemmons and R. Buchholz*
In *Behavioral Approaches to Conservation in the Wild*, 1997
Genetic diversity and behavioral diversity are both important to the survival of species.

We have to be concerned, in short, with the total quality of the environment.
— *Laurance S. Rockefeller*
In a report to President Lyndon B. Johnson, May 1965

I am I plus my surroundings and if I do not preserve the latter, I do not preserve myself.
— *Jose Ortega y Gasset*
In *Meditations on Quixote*, 1914

There is nothing more practical in the end, than the preservation of beauty, than the preservation of anything that appeals to the higher emotions of man.
— *President Theodore Roosevelt*
Ecological importance, not beauty, should determine what we preserve.

Communities should be planned with an eye to the effect made upon the human spirit by being continuously surrounded with a maximum of beauty.
— *President Thomas Jefferson*

There is no doubt about it, the basic satisfaction in farming is manure, that always suggests that life can be cyclic and chemically perfect and aromatic and continuous.
— *E.B. White*

Is a park any better than a coal mine? What's a mountain got that a slag pile hasn't? What would you rather have in your garden—an almond tree or an oil well?
— *Jean Giraudoux*
In *The Madwoman of Chaillot*, spoken by a prospector

Is it possible that I am not alone in believing that in the dispute between Galileo and the Church, the Church was right and the center of *man's* universe is the earth?
— *Stephen Vizinczey*
In *Rules of the Game*, 1970

If you besiege a town for a long time, making war against it in order to take it, you must not destroy its trees by wielding an ax against them. Although you may take food from them, you must not cut them down.
> — *Deuteronomy 20:19-20*

Anyone who benefits from the world without saying a blessing is a thief.
> — *Babylonian Talmud, Berachot 35a*

A world that is no longer fit for wild plants, for the flight of birds, a world whose streams and forests are empty and lifeless is not likely to be fit habitat for man himself, for these things are symptoms of an ailing world.
> — *Rachel Carson*
> Quoted in the *Audubon Magazine*

Nature-faker.
> — *President Theodore Roosevelt*
> In *Everybody's Magazine*, September, 1907
> *In reference to those who pretend to be conservationists.*

The "control of nature" is a phrase conceived in arrogance.
> — *Rachel Carson*

We abuse land because we regard it as a commodity belonging to us. When we see land as a community to which we belong, we may begin to use it with love and respect.
> — *Aldo Leopold*
> In *A Sand County Almanac*, 1949

Everybody likes to hear about a man laying down his life for his country, but nobody wants to hear about a country giving her shirt for her planet.
> — *E. B. White*
> In "World Government and Peace" in *The New Yorker*, 1943–1945

There are hidden contradictions in the minds of people who "love Nature" while deploring the "artificialities" with which "Man has spoiled 'Nature.'" The obvious contradiction lies in their choice of words, which imply that Man and his artifacts are *not* part of "Nature"—but beavers and their dams *are*.
> — *Robert Heinlein*
> In *Time Enough for Love*, 1973

Restoration

The American people have a right to air that they and their children can breathe without fear.
—*President Lyndon B. Johnson*

Any nation concerned about the quality of life, now and forever, must be concerned about conservation. It will not be enough to merely halt the damage we've done. Our national heritage must be recovered and restored.
— *George Bush, 1988*

We must not only protect the countryside and save it from destruction, we must restore what has been destroyed and salvage the beauty and charm of our cities.
— *President Lyndon B. Johnson*
In his "Message on the Natural Beauty of Our Country," February 8, 1965

Renewal is the way of nature, and it must now become the way of man.
— *George Bush, 1988*

Saving the planet has never been an issue of money, but rather a matter of resourcefulness and motivation of individuals.
— *Spencer Beebe*

...the American people are awakening to the new concept that the river basins are economic units; that many problems center around the use and control of the water resources.
— *President Harry Truman's Water Policy Commission, 1950*

The perils of planet Earth—we read about them, we brood about them, but let's admit it, often as not we try to ignore them.
—*Janet Marinelli*

The significant problems we face cannot be solved at the same level of thinking we were at when we created them.
— *Albert Einstein*
Now that we have elevated our level of thinking, perhaps we can solve some of the environmental problems that were created in years past.

Efforts to "improve the quality of stream fishing" usually focus on altering physical habitat, as if physical habitat were a magic wand controlling fish abundance and distribution.

> — *Charles F. Rabeni and Scott P. Sowa*
> In *Integrating Biological Realism into Habitat Restoration and Conservation Strategies for Small Streams*

A thing is right when it preserves the integrity, stability, and beauty of the biotic community. It is wrong when it tends otherwise.

> — *Aldo Leopold*
> In *A Sand Country Almanac*, 1949
> *This statement is often referred to as "The Land Ethic."*

Conservation is a state of harmony between man and land.

> — *Aldo Leopold*
> In the essay "Conservation," 1953

First, do no harm.

> — *Hippocratic Oath*
> *The principle ethic of physicians is equally applicable to the well-meaning efforts of environmental restoration.*

American conservation is, I fear, still concerned for the most part with showpieces. We have not yet learned to think in terms of small cogs and wheels.

> — *Aldo Leopold*
> In the essay "Conservation Ethic," 1938

There is a phenomenal resiliency in the mechanisms of the earth. A river or lake is almost never dead. If you give it the slightest chance by stopping the pollutants going into it, then nature usually comes back.

> — *Rene Dubos, 1981*

With a fork drive Nature out, she will ever yet return.

> — *Horace, 20 B.C.E.*
> In *Epistles, I*. Quoted and translated in "Marigolds" by Robert Graves.
> *Originally, "Naturam expelles furca, tamen usque recurret."*

He who plants a tree plants hope.

> — *Lucy Larcom*
> In "Plant a Tree"
> *Larcom also said, "He who plants a tree plants a joy."*

The poetry of the earth is never dead.
—*John Keats*

I know it when I see it.
— *Potter Stewart, Supreme Court Justice*
Referring to pornography but is equally applicable to the difficulty in providing guidelines on environmental preservation and restoration.

It is axiomatic that no restoration can ever be perfect; it is impossible to replicate the biogeochemical and climatological sequence of events over geological time that led to the creation and placement of even one particle of soil, much less to exactly reproduce an entire ecosystem. Therefore, all restorations are exercises in approximation and in the reconstruction of naturalistic rather than natural assemblages of plants and animals with their physical environments.
— *Berger, 1990*

One's philosophy is not best expressed in words, it's expressed in the choices one makes. In the long run, we shape our lives and we shape ourselves. The process never ends until we die. And the choices we make are ultimately our responsibility.
— *Eleanor Roosevelt*

A long, healthy, and happy life is the result of making contributions, of having meaningful projects that are personally exciting and contribute to and bless the lives of others.
— *Hans Selye*

Never tear a building down from the bottom up.
— *American proverb*

Painted flowers have no scent.
— *American proverb*

It is the continuing policy of the Federal government to foster and promote the general welfare, to create and maintain conditions under which man and nature can exist in productive harmony, and to fulfill the social, economic, and other requirements of present and future generations of America.
— *National Environmental Policy Act (NEPA), section 101, 1969*

It became necessary to destroy the town in order to save it.

— *U.S. Army officer*

Spoken in reference to the Vietnamese town of Ben Tre.

Power of Nature

Civilization exists by geological consent, subject to change without notice.

— *Will Durant*

Quoted in *Archaeology* by David Thomas

That which is not good for the bee-hive cannot be good for the bees.

— *Marcus Aurelius*

In *Meditations*

We have a major disaster developing.

— *Emergency Operations Center of the Army Corps of Engineers, 1972*

Quoted in "The Agnes Disaster" by Maj. Gen. Richard H. Groves in *The Military Engineer*

Spoken on the eve of flooding caused by Tropical Storm Agnes. The floods caused $3.1 billion in damages, killed 122 people, and destroyed 115,000 buildings.

Who has cut a channel for the torrents of rain, and a way for the thunderbolt, to bring rain on a land where no one lives, on the desert, which is empty of human life, to satisfy the waste and desolate land, and to make the ground put forth grass?

— *Job 38:25-27*

Without the natural elements, life cannot exist.

Man cannot prevent these floods which are bound to recur but he can, to a great extent, take measures to minimize their effects.

— *The Floods of March 1936 in Pennsylvania*

Prepared by the Commonwealth of Pennsylvania Department of Forests and Waters in cooperation with the United States Geological Survey, 1936.

The floods of March 1936 reportedly resulted in over $212 million in damages, 80 lives lost, 2,822 injuries and 57,800 homes damaged or destroyed in Pennsylvania alone.

…in the spring of 1935, the sun darkened from the Rocky Mountains to the Atlantic by vast clouds of soil particles borne from the denuded grasslands of the Western States.
— *Fairfield Osborn*
Quoted in *From Sea to Shining Sea* by The Presidents Council on Recreation and Natural Beauty

Once more the elemental forces of nature had conquered—but such victories are only temporary. After each such setback, man proceeds with more perfect knowledge, with greater resourcefulness, and with strengthened resolve, to strive again, to play again, to build again, to achieve again—all toward renewed triumph and more enduring mastery over the obstacles and destructive forces of nature.
— *David B. Steinman*
Quoted in *The Book of Bridges* by M. Hayden
Bridge engineer credited with greatly increasing the understanding of aerodynamic bridge design. The most notable failure due to wind-induced vibrations was the Tacoma Narrows Bridge over Puget Sound which failed in 25–40 mph winds in 1940.

One dark and stormy night, a long time ago, a sailing warship made its way through heavy seas. Suddenly, a light appeared out of the gloom in its path.
  The captain of the warship called out, "Unknown vessel, change your course to starboard."
  A voice came back, "No, you must turn to starboard."
  The captain began to get angry and shouted, "You are in our path. Turn to starboard immediately!"
  The voice came back, "You change your course to starboard. Please comply!"
  The captain began to fume and yelled, "This is a captain you are speaking to! Change your course immediately!"
  The voice came back, "This is seaman first class Jones. You change your course to starboard immediately!"
  The captain stomped his feet and yelled, "This is a warship on official business. Change your course to starboard or be destroyed!"
  The voice came back, "Sir, this is a lighthouse!"
— *Anonymous*
Moral: There are some forces that cannot be altered.

There is a tendency to look upon a great flood as unprecedented and not likely to occur again. As a matter of fact great floods at long intervals of time are as much a part of the normal course of events as are smaller ones which occur at much more frequent intervals.

— *The Floods of March 1936 in Pennsylvania*
Prepared by the Commonwealth of Pennsylvania Department of Forests and Waters in cooperation with the United States Geological Survey, 1936.

Sustainable Development

Our ideals, laws and customs should be based on the proposition that each generation in turn becomes the custodian rather than the absolute owner of our resources—and each generation has the obligation to pass this inheritance on to the future.

— *Alden Whitmann*

Humanity has the ability to make development sustainable—to ensure that it meets needs of the present without compromising the ability of future generations to meet their own needs.

— *The World Commission on Environment and Development*
In *Our Common Future,* 1987

The stationary state would make fewer demands on our environmental resources, but much greater demands on our moral resources.

— *Herman Daly*
In "Toward a Stationary-State Economy" in *Patient Earth* by J. Harte and R. Socolow, 1971

Today I am more than ever frightened. I wish it would dawn upon engineers that, in order to be an engineer, it is not enough to be an engineer.

— *Jose Ortega y Gasset*
Engineers must see how their work can have influence far beyond the one project upon which they are working.

Wilderness is the raw material out of which man has hammered the artifact called civilization.

— *Aldo Leopold*
In *A Sand County Almanac,* 1949

If I were a Brazilian without land or money or the means to feed my children, I would be burning the rain forest too.

> — *Gordon Matthew Sumner, a.k.a. "Sting"*
> In the *International Herald Tribune*, April 14, 1989
>
> *Environmental concerns and economic concerns, although inherently connected, are not necessary mutually exclusive. The Brazilian must be able to feed his children before he can worry about his children's children.*

Growth for the sake of growth is the ideology of the cancer cell.

> — *Edward Abbey*

Conservation must come before recreation.

> — *Charles, Prince of Wales*
> In the *Times* of London, July 5, 1989

Many present efforts to guard and maintain human progress, to meet human needs, and to realize human ambitions are simply unsustainable—in both of the rich and poor nations. They draw too heavily, too quickly, on already overdrawn environmental resources accounts. ... They may show profits on the balance sheets of our generation, but our children will inherit the losses.

> — *The World Commission on Environment and Development*
> In *Our Common Future*, 1987

The future is no longer what it was thought to be, or what it might have been if humans had known how to use their brains and their opportunities more effectively. But the future can still become what we reasonably and realistically want.

> —*Aurelio Peccei*
> In *One Hundred Pages for the Future: Reflections of the President of the Club of Rome*, 1981

Biological diversity is the key to the maintenance of the world as we know it. Life in a local site struck down by passing storm springs back quickly: opportunistic species rush in to fill the spaces. They entrain the succession that circles back to something resembling the original state of the environment.

> — *Edward O. Wilson*
> In *The Diversity of Life*, 1992

Merely the attempt to solve the biodiversity crisis offers great benefits never before enjoyed, for to save species is to study them closely, and to learn them well is to exploit their characteristics in novel ways.

— *Edward O. Wilson*
In *The Diversity of Life*, 1992

Faced with the widespread destruction of the environment, people everywhere are coming to understand that we cannot continue to use the goods of the earth as we have in the past...a new ecological awareness is beginning to emerge which rather than being downplayed, ought to be encouraged to develop into concrete programs and initiatives.

— *Pope John Paul II, December 8, 1989*
Quoted by Al Gore in *Earth in the Balance—Ecology and the Human Spirit*

All the evidence suggests that we have consistently exaggerated the contributions of technological genius and underestimated the contributions of natural resources....We need ... something we lost in our haste to make the world: a sense of limits, an awareness of the importance of earth's resources.

— *Stewart Udall*
Quoted in *Overshoot: The Ecological Basis of Revolutionary Change* by William R. Catton, Jr.

If the federal government had been around when the Creator was putting His hand to this state, Indiana wouldn't be here. It'd still be waiting for an environmental impact statement.

— *Ronald Reagan*
In a speech on February 9, 1982

Can we move nations and people in the direction of sustainability? Such a move would be a modification of society comparable in scale to only two other changes: the Agricultural Revolution of the late Neolithic and the Industrial Revolution of the past two centuries. Those revolutions were gradual, spontaneous, and largely unconscious. This one will have to be a fully conscious operation, guided by the best foresight that science can provide.... If we actually do it, the undertaking will be absolutely unique in humanity's stay on Earth.

— *William D. Ruckelshaus*
In *Toward a Sustainable World* in *Scientific American*, 1989

Chapter 5
Speaking of Science
Nature

The world makes a messy laboratory for ecologists.
— *James Gleick*
In *Chaos: Making a New Science*, 1987

The most incomprehensible thing about the world is that it is at all comprehensible.
— *Albert Einstein*

As I explained in the first lecture, the way we have to describe Nature is generally incomprehensible to us.
— *Richard Feynman*
In *QED: The Strange Theory of Light and Matter*, 1990

Thou, nature, art my goddess; to thy laws my services are bound.
— *William Shakespeare*
In *King Lear*, 1605
This was one of the mottoes for the great physicist and mathematician Karl Gauss.

My father considered a walk among the mountains as the equivalent of churchgoing.
— *Aldous Huxley*

Come forth into the light of things. Let Nature be your teacher.
— *William Wordsworth*

The only solid piece of scientific truth about which I feel totally confident is that we are profoundly ignorant about nature.
— *Lewis Thomas*
In *The Medusa and the Snail*, 1979

Nature never breaks her own laws.
— *Leonardo da Vinci*

Nature is what wins in the end.
— *Abby Adams*
In the *Gardener's Gripe Book*, 1995

Nature uses as little as possible of anything.
—*Johannes Kepler*

My sun sets to rise again.
— *Robert Browning*

It is not the language of the painters but the language of nature to which one has to listen.
— *Vincent van Gogh*

Nature is a mutable cloud which is always and never the same.
— *Ralph Waldo Emerson*
In *Essays: First Series*, 1841

Nature reserves some of her choice rewards for days when her mood may appear to be somber.
— *Rachel Carson*
In *The Sense of Wonder*, 1956
In other words, some of nature's prettiest days àre, in fact, rainy days.

There is symbolic as well as actual beauty in the migration of the birds, the ebb and flow of the tides, the folded bud ready for the spring. There is something infinitely healing in the repeated refrains of nature—the assurance that dawn comes after night, and spring after the winter.
— *Rachel Carson*
In *The Sense of Wonder*, 1956

April hath put a spirit of youth in every thing.
— *William Shakespeare*

To live is so startling it leaves little time for anything else.
— *Emily Dickinson*

Remember what you have seen, because everything forgotten returns to the circling winds.
— *Navajo Wind Chant*

Adopt the pace of nature; her secret is patience.
— *Ralph Waldo Emerson*

Nature is but an image or imitation of wisdom, the last thing of the soul;
nature being a thing which doth only do, but not know.
— *Plotinus*

Nature is the art of God.
— *Dante Alighieri*

As it is written in the book of the words of Isaiah the prophet, saying,
The voice of one crying in the wilderness, Prepare the way for the Lord,
make his paths straight. Every valley shall be filled, and every mountain
and hill shall be brought low; and the crooked shall be made straight,
and the rough ways shall be made smooth.
— *Luke 3:4-5*
But the environmentalists would have a fit!

These are the Gardens of the Desert, these the unshorn fields, boundless
and beautiful, and fresh as the young earth, ere man had sinned—
the prairies.
— *William Cullen Bryant*
In *The Prairies*, 1832

There is a pleasure in the pathless woods,
There is a rapture on the lonely shore,
There is society, where none intrudes,
By the deep Sea, and the music in its roar:
I love not Man the less, but Nature more.
— *George Gordon Byron*
In *Childe Harold's Solitude*, 1812-1818

Nature is usually wrong.
— *James Whistler*

All art is but imitation of nature.
— *Seneca*
In *Epistles*, 63 C.E.

Nature is honest, we aren't; we embalm our dead.
— *Ugo Betti*

I don't understand those who equate the *All Natural* label with the product automatically being healthy. For example, ask Socrates about the health benefits of all natural hemlock.
— *Archie Fripp, Jr.*
Hemlock is the deadly poison that killed Socrates.

Plants

If a tree dies, plant another in its place.
— *Carolus Linnaeus*

Next in profusion to the divine profusion of water, light and air, those three physical facts which render existence possible, may be reckoned the universal beneficence of grass.
— *John James Ingalls*
In *Blue Grass*

As far as naturalists are concerned, the great value of wildflowers is in perpetuating the species, in contributing their very presence to the environment in which they play an integral part.
— *Barbara Burn*
In *North American Wildflowers*

Convince me that you have a seed there, and I am prepared to expect wonders.
— *Henry David Thoreau*

The way botanists divide up flowers reminds me of the way Africa was divided into countries by politicians.
— *Miles Kington*
In *Nature Made Ridiculously Simple*, 1983

The maples and ferns are still uncorrupt, Yet, no doubt, when they come to consciousness, they too, will curse and swear.
— *Ralph Waldo Emerson*
In *Nature*, 1836

When I touch a flower, I am touching Infinity. Flowers existed long before there were human beings on this earth, and they will continue to exist for millions of years after.
— *George Washington Carver*

What is a weed? A plant whose virtues have not yet been discovered.
— *Ralph Waldo Emerson*
In *Fortune of the Republic*, 1878

A weed is no more than a flower in disguise.
— *Ella Wheeler Wilcox*
In *The Weed*

Nature writes, gardeners edit.
— *Roger Swain*
In *Greenprints*

More than anything I must have flowers, always, always.
— *Claude Monet*

One of nature's loveliest sights is a field in springtime filled with colorful flowers, creating patterns of blue, lavender, yellow, pink, and white as far as the eye can see.
— *Barbara Burn*
In *North American Wildflowers*

Every flower is a soul blossoming in nature.
— *Gerard De Nerval*

Life begins the day you start a garden.
— *Chinese proverb*

Earth laughs in flowers.
— *Ralph Waldo Emerson*

One for the blackbird, one for the crow, one for the cutworm, and one to grow.
— *Gardening proverb for planting seeds.*

Sometimes people as well as trees need careful pruning.
— *Barbara Rosen*

Before the seed there comes the thought of bloom.
— *E.B. White*

It is not enough for a gardener to love flowers; he must also hate weeds.
— *American proverb*

If a person cannot love a plant after he has pruned it, then he has either done a poor job or is devoid of emotion.
— *Liberty Hyde Bailey*

One who plants a garden, plants happiness.
— *Chinese proverb*

The very earth itself is a granary.
— *Henry David Thoreau*
    A granary is a storehouse for grain.

We're proud of humanity's powers
But these potions and medicines of ours
Coffee, garlic, and spices,
Evolved as devices
So that insects would stop bugging flowers.
— *Richard Cowen*
    In *History of Life*, 1989

Forests

Without enough wilderness America will change. Democracy, with its myriad personalities and increasing sophistication, must be fibered and vitalized by the regular contact with outdoor growths—animals, trees, sun warmth, and free skies—or it will dwindle and pale.
— *Walt Whitman*

Keep a green tree in your heart, and perhaps the singing bird will come.
— *Chinese proverb*

I went to the woods because I wished to live deliberately.
— *Henry David Thoreau*

When you're alone in the forest, you're aware that life is everywhere around you.
— *Russell Mittermeir*
Quoted in *Time Magazine*, 1998
Primatologist speaking about rain forests.

Groves were God's first Temples.
— *William Cullen Bryant*

Trees are the earth's endless effort to speak to the listening heaven.
— *Rabindranath Tagore*

I like trees because they seem more resigned to the way they have to live than other things do.
— *Willa Cather*
In *O Pioneers!*, 1913

You can't chop down a forest without a few splinters flying.
— *Russian proverb*
Quoted in *Sovietski Collection*

A good tree cannot bring forth evil fruit, neither can a corrupt tree bring forth good fruit.
— *Matthew 7:18*

I think that I shall never see a poem as lovely as a tree.
> —*Joyce Kilmer*
> In *Trees*, 1914

Here man is no longer the center of the world, only a witness, but a witness who is also a partner in the silent life of nature, bound by secret affinities to the trees.
> — *Dag Hammarskjold*

Wilderness is a resource which can shrink but not grow.
> — *Aldo Leopold*
> In *A Sand County Almanac*, 1949
> *Fortunately, this sentiment is overly pessimistic. For example, 75% of New England is covered by forests compared with only 20% a century ago.*

Every leaf speaks bliss to me.
> — *Emily Bronte*

Wetlands

...the horrible desert, the foul damps ascend without ceasing, corrupt the air and render it unfit for respiration...never was rum—that cordial of life—found more necessary than in this dirty place.
> — *Colonel William Byrd, III*
> *Written in the 17th century while in the Great Dismal Swamp of Virginia.*

...swamps and overflowed land, which may be or are found unfit for cultivation.
> — *Swamp Land Act, 1849*
> *Description of wetlands. The Swamp Land Act was the nation's first federal wetland policy. It granted the state of Louisiana jurisdiction over wetlands for control of Mississippi Basin flooding. By 1860, 14 other states were included and 64.9 million acres of wetlands were granted to these 15 states in exchange for promises to drain and convert them to farmland.*

Wetlands hold seemingly magical properties that make them unique among major ecosystem groups in the world.
> — *R. H. Kadlec and R. L. Knight*
> In *Treatment of Wetlands*

Wetlands are an irreplaceable natural resource which in its natural state is essential to the ecological systems of the tidal rivers, bays, and estuaries of the Commonwealth.
> — *Virginia Wetland Act, 1972*
>> *Compare how the attitudes towards swamps and wetlands have evolved over the past centuries by comparing this quote to the previous two.*

Reeds will bend, but iron will not.
> — *American proverb*

Where there are reeds, there is water.
> — *American proverb*

There is no wilder and richer sight than is afforded from such a point of view as of the edge of a blueberry swamp.
> — *Henry David Thoreau*

…a charming field for an encounter.
> — *George Washington, May 24, 1754*
>> *In reference to the field that he selected to build Fort Necessity at the beginning of the French and Indian War. On July 3rd, approximately 700 French and Indians attacked George Washington's force of about 400. Concurrently, it began to rain, which flooded the low marshy ground. Washington soon found his powder wet and his troops becoming ill in the trenches and creeks where he had placed them. Within a few hours, Washington surrendered. Many of the calamities in modern times which are a result of construction in wetlands are of a legal nature and far less bloody.*

Rivers

Human activity has profoundly affected rivers and stream in all parts of the world, to such an extent that it is now extremely difficult to find any stream which has not been in some way altered, and probably quite impossible to find any such river.
> — *H.B.N. Hynes, 1970*

The care of rivers is not a question of rivers, but of human heart.
— *Tanka Shozo*

Crooked streams are a menace to life and crops…The twisting and
turning of the channel retards flow and reduces the capacity of the
stream to handle large volumes of water. Floods result. …take the kinks
out of crooked streams…Dupont dynamite has straightened many
thousands of miles of crooked streams.
— *1935 Dupont dynamite advertisement in* American Forests Journal
Quoted in *Applied River Morphology* by Dave Rosgen

Most rivers if left completely to their own devices, will develop qualities
that are at best inconvenient to man and his activities and at worst a
threat to life and property.
— *F.M. Henderson*
In *Open Channel Flow*, 1966

To protect your rivers, protect your mountains.
— *attributed to Emperor Yu of China, 2000 B.C.E.*
Quoted in *Records of the Historians* by Sima Qian, circa 100 B.C.E.
*It is reported that Yu felt such responsibility to his flood-control work that
he left home just four days after his wedding to return to his job. It appears
that the personality of engineers has not changed dramatically over four
thousand years.*

The ecological integrity of river ecosystems depends on their natural
dynamic character.
— *N.L. Poff, J.D. Allen, M.B. Bain, J.R. Karr, K.L. Prestegaard, B.D. Richter,
R.E. Sparks, and J.C. Stromberg*
In "The Natural Flow Regime" in *Bioscience*, 1997

Rivers and the Inhabitants of the watery Element were made for wise
men to contemplate and fools to pass by.
— *Izaak Walton*
In *The Compleat Angler*, 1653

You can not step twice into the same river, for other waters are
continually flowing on… It is in changing that things find repose.
— *Heraclitus*
In *Fragments* 21, 23

The River is my own highway, the only wild and unfenced part of the world.
— *Henry David Thoreau, 1852*

Too thick to drink and too thin to plow.
— *Anonymous, 1800s*
  Derisive comment used to describe the high sediment in the Mississippi. Bank erosion has always been significant on the Mississippi. For example, between 1855 and 1860, the entire business district of Grand Gulf, MS, 55 city blocks, was lost to bank erosion.

The St. Lawrence is water, and the Mississippi is muddy water; but that, sir, is liquid history.
— *John Burns*
  Said on the terrace of the House of Commons in London in reply to trans-atlantic visitors who belittled the size of the Thames River.

While floods might be considered acts of God, flood losses are the results of acts of man. The problem thus becomes one of adjusting the human habitat in the flood plain environment with effective resource management.
— *Gilbert F. White*
  In *The Flood Control Challenge: Past, Present, and Future,* 1942

If we change a river we usually do some good somewhere and "good" in quotation marks. That means we achieve some kind of a result that we are aiming at but sometimes forget that the same change which we are introducing may have widespread influences somewhere else. I think if, out of today's emphasis of the environment, anything results for us it is that it emphasizes the fact that we must look at a river or a drainage basin or whatever we are talking about as a big unit with many facets.
— *Hans Albert Einstein*
  In the Army Corp of Engineers' Waterways Experimental Station's *Streambank Stabilization Handbook,* 1997

Humans have long been fascinated by the dynamism of free-flowing waters. Yet we have expended great effort to tame rivers for transportation, water supply, flood control, agriculture, and power generation.
— *N.L. Poff, J.D. Allen, M.B. Bain, J.R. Karr, K.L. Prestegaard, B.D. Richter, R.E. Sparks, and J.C. Stromberg*
  In "The Natural Flow Regime" in *Bioscience,* 1997

No stream rises higher than its source.
> — *Frank Lloyd Wright*

Only the game fish swims up stream, but the sensible fish swims down.
> — *Grantland Rice*
> In *The Ballade of the Gamefish*

A dead fish can float downstream, but it takes a live one to swim upstream.
> — *American proverb*

Rivers know this: there is no hurry. We shall get there some day.
> — *Joan Powers*
> In *Pooh's Little Instruction Book*

The man who is swimming against the stream knows the strength of it.
> — *Woodrow Wilson*

All the rivers run into the sea; yet the sea is not full; unto the place from which the rivers come, thither they return again.
> — *Ecclesiastes 1:7*
>> *There has not been much improvement on the basic description of the water cycle until this century when the multiple paths of the water cycle were quantified.*

The sea is the source of the waters and the source of the winds. Without the great sea, not from the clouds could come the flowing rivers or the heaven's rain.
> — *Xenophanes of Colophon, circa 550 B.C.E.*
> Quoted in *Land, Water and Development* by M. Newson
> *This is an early description of the water cycle.*

And like the ocean, day by day receiving Floods from all lands, which never over flows; Its boundary-line not leaping, and not leaving, Fed by rivers, but unswelled by those.
> — In the *Bhagavadgita*, circa 100 C.E., spoken by Krishna
>> *Descriptions of the water cycle can be found in nearly every civilization, including this quote from the Hindu epic poem.*

Egypt is the gift of the river.
— *Herodotus*

Water is so much more fine and sensitive an element than earth.
A single boatman passing up or down unavoidably shakes the whole of
a wide river.
— *Henry David Thoreau*
In a journal entry, September 19, 1850

Without being the owner of any land, I find that I have a civil right to
the river.
— *Henry David Thoreau*
In a journal entry, March 23, 1850

And he shall be like a tree planted by the rivers of water, that brings
forth his fruit in his season; his leaf also shall not wither; and whatsoever
he does shall prosper.
— *Psalms 1:3*

And convey good news to those who believe and do good deeds, that
they shall have gardens in which rivers flow.
— *The Koran 2:25*

I do not know much about gods; but I think that the river
Is a strong brown god—sullen, untamed, and intractable.
— *Thomas Stearns Eliot*
In *Four Quartets: The Dry Salvages*, 1943

The depth of a fordable river or stream is no deterrent if you can keep
your footing.
— *U.S. Army Survival Manual, FM 21-76, 1998*

The Missouri was a river to make strong men weep and rich men poor.
— *Richard Edward Oglesby*

Oceans

The use of the sea and air is common to all; neither can a title to the ocean belong to any people or private persons, forasmuch as neither nature nor public use and custom permit any possession thereof.
> — *Elizabeth I, Queen of England*
> In a letter to the Spanish Ambassador, 1580

In the biting honesty of salt, the sea makes her secrets known to those who care to listen.
> — *Sandra Benitez*
> In *A Place Where the Sea Remembers,* 1993

If you want to know all about the sea…and ask the sea itself, what does it say? Grumble grumble swish swish. It is too busy being itself to know anything about itself.
> — *Ursula K. Le Guin*
> In "Talking about Writing" in her book *Language of the Night,* 1979

Roll on, thou deep and dark blue ocean—roll!
> — *George Gordon Byron*
> In *Childe Harold's Pilgrimage,* 1812–1818

Unchangeable, save to thy wild waves play.
> — *George Gordon Byron*
> In *Childe Harold's Pilgrimage,* 1812–1818

If the sea is sick, we'll feel it. If it dies, we die. Our future and the state of the oceans are one.
> — *Sylvia A. Earle*
> In *Sea Change,* 1995

My soul roams with the sea.
> — *"The Seafarer"*
> An Anglo-Saxon poem translated by Burton Raffel

Go down to the sea in ships,… do business in great waters.
> — *Psalms 107:23*

The mysterious human bond with the great seas that poets write about
has a physiological base in our veins and in every living thing, where
runs fluid of the same saline proportions as ocean water.

> — *Anne W. Simon*
> In *The Thin Edge*, 1978

It doesn't matter where on Earth you live, everyone is utterly dependent
on the existence of that lovely, living saltwater soup. There's plenty of
water in the universe without life, but nowhere is there life without water.

> — *Sylvia A. Earle*
> In *Sea Change*, 1995

On the reverse of the 1993 German 10 mark note which honors
the great mathematician Carl Gaus is a picture of a sextant.
The sextant, invented by John Campbell in 1757, allowed sailing
ships to approximate their position.

Men go back to the mountains, as they go back to sailing ships at sea,
because in the mountains and on the sea they must face up, as did men
of another age, to the challenge of nature. Modern man lives in a highly
synthetic kind of existence. He specializes in this and that. Rarely does
he test all his powers or find himself whole. But in the hills and on the
water the character of a man comes out.

> — *Abram T. Collier*

Give her to the God of Storms, the lightning and the gale.

> — *Oliver Wendell Holmes*
> In the poem *Old Ironsides*, 1830
> *Written in protest of the U.S.S. Constitution being sold by the U.S. Navy.*
> *The ship has remained in active commission of the U.S. Navy.*

There is witchery in the sea, its songs and stories…
> — *Richard Henry Dana Jr.*
> In *Two Years Before the Mast*, 1840

The boisterous winds and neptune's waves have tost me too and fro,
by Gods decree you plainly see I am harbour'd here below.
> — *1768 tombstone of mariner James Barnerd buried in the*
> *Hudson Valley*
> Quoted in *Sail Magazine*

The whole ocean is made up of little drops.
> — *American proverb*

Praise the sea, but keep on land.
> — *American proverb*

I like rivers better than oceans, for we see both sides. An ocean is forever
asking questions and writing them aloud along the shore.
> — *Edwin Arlington Robinson*
> In *Roman Bartholow*

Whales and Dolphins

The tempests fly before their father's face,
Trains of inferior gods his triumph grace,
And monster whales before their master play,
And choirs of Tritons crowd the wat'ry way.
> — *Virgil*
> In *The Aeneid*, 29–19 B.C.E.

Among the sea-fishes many stories are told about the dolphin, indicative
of his gentle and kindly nature, and of manifestations of passionate
attachment to boys…. It appears to be the fleetest of all animals, marine
and terrestrial, and it can leap over the masts of large vessels.
> — *Aristotle*
> In the *History of Animals*, 350 B.C.E.
>
> *Aristotle continues by detailing dolphin behaviors. Some of these behaviors*
> *have been systematically studied in this century. Although dolphins often*
> *appear "gentle and kindly," they can also be aggressive and dangerous.*

A cross between a sea-serpent and an alligator.
> — *Charles Melville Scammon*
> In *The Marine Mammals of the Northwestern Coast of North America*, 1874
> *The whalers' description of the California gray whale.*

Cetacean [whale and dolphin] sounds and their roles in the animals' lives will continue to inspire creative speculation and useful research. Certainly they are as eerie a reminder of the mystery of the sea as can be found.
> — *Stephen Leatherwood and Randall Reeves*
> In *The Sierra Club Handbook of Whales and Dolphins*, 1983

It is hardly necessary to say, that any person taking up the study of marine mammals, and especially the Cetaceans, enters a difficult field of research, since the opportunities for observing the habits of these animals under favorable conditions are but rare and brief.
> — *Charles Melville Scammon*
> In *The Marine Mammals of the Northwestern Coast of North America*, 1874

It is obvious that no matter where and how it is studied, the whale requires the application of a wide range of innovative methodologies and techniques.
> — *Howard E. Winn and Bori L. Olla*
> In *Behavior of Marine Mammals: Current Perspectives in Research*, 1979

A full understanding of the behavior of marine mammals requires studies both at sea and in captivity.
> — *Ken Norris*
> In a talk entitled "The Use of Captive Marine Mammals in Behavior Studies," 1985

Before the invention of the harpoon gun, whalers were forced to study the peculiarities of the whale's behavior in more detail and more scrupulously not only for the success of the whaling but also for their own safety.
> — *Alexey V. Yablokov*
> Quoted in *The Sperm Whale* by A. A. Berzin

Intelligent as some of our whalemen have been … it must be borne in mind that their main object is the capture of these valuable prizes.
> — *T. Southwell*
> In *The Migration of the Right Whale,* 1898

Navy training films portrayed killer whales as dangerous vermin that might attack lifeboats and swimmers; some military fliers reportedly used them for bombing practice, in the belief that they were thus protecting their buddies.
> — *Karen Pryor and Ken Norris*
> In "Dolphin Politics and Dolphin Science" in *Dolphin Societies,* 1991
> *Fortunately, the public impression of killer whales has changed radically since the WWII-era films.*

Aquariums can take credit for first bringing dolphins and whales to the world's attention as remarkable mammals that have family life and social behavior analogous to other mammals. Before this, these animals were seen merely as sources of meat, oil, and leather products.
> — *Murray A. Newman*
> In *Life in a Fishbowl,* 1994

Perhaps instead of thinking of whales in terms of aggregates, we should think about them as individuals operating in a social context that is maintained by complex individual social interactions.
> — *G. Bartholemew*
> In "The Relation of the Natural History of Whales to Their Management" in *The Whale Problem: a Status Report* by Schevill, Ray, & Norris, 1974
> *One of the biggest impediments to the conservation of whales has been the mindset that whales can be thought of as units of oil, rather than as individual animals.*

If you were to make little fishes talk, they would talk like whales.
> — *James Boswell*
> In the *Life of Johnson,* 1785

I invite you to entertain some new beliefs about dolphins … [that] *these Cetacea with huge brains are more intelligent than any man or woman.*
> — *John C. Lilly*
> In *Communication between Man and Dolphin: The Possibilities of Talking with Other Species,* 1978
> *There is no evidence for Lilly's claim. Current research suggests a dolphin's intelligence is more similar to a chimpanzee's than to a human's intelligence.*

Individual dolphins and whales are to be given the legal rights of human individuals ... Research into communication with cetaceans is no longer simply a scientific pursuit ... We must learn their needs, their ethics, their philosophy, to find out who we are on this planet, in this galaxy. The extraterrestrials are here—in the sea.

—*John C. Lilly*
In *The Rights of Cetaceans under Human Laws*, 1976

This statement has been summarized by Edward O. Wilson: "Lilly's writing differs from that of Herman Melville and Jules Verne not just in its more modest literary merit but more basically in its humorless and quite unjustified claim to be a valid scientific report."

Don't tell fish stories where the people know you; but particularly, don't tell them where they know the fish.

— *Mark Twain*

It is of interest to note that while some dolphins are reported to have learned English—up to fifty words used in correct context—no human being has been reported to have learned dolphinese.

— *Carl Sagan*

While dolphins have been taught to recognize words and their context, no dolphin has ever learned how to use words in context. Similarly, there is no evidence that there is any form of "dolphinese" or any other form of language in dolphins.

We found no evidence indicating a "song patterning" or "language." The level of information content in the whistles may ... even exceed that of other advanced social animals but is much inferior in specificity to even a rudimentary language.

— *Melba Caldwell, David Caldwell, and R. Turner*
In *Statistical Analysis of the Signature Whistle of an Atlantic Bottle-nose Dolphin with Correlation between Vocal Changes and Level of Arousal*, 1970

Complex dolphin sociology and high cetacean intelligence have joined motherhood and apple pie in the public mythology.

— *D. E. Gaskin*
In *The Ecology of Whales and Dolphins*, 1982

One of the popular myths is that dolphins commonly save swimmers by pushing them to shore. While this has undoubtedly occurred, dolphin behavior is such that an equal number of swimmers will be pushed away from shore because the dolphins find the swimmer an amusing toy with which to play.

Can you be a serious scientist if you work with dolphins?

— *Ken Norris*
In "Looking at Captive Dolphins" in *Dolphin Societies,* 1991
This statement summarizes the scientific community's reaction to the anthropomorphization of dolphins.

Admiral, there be whales here!

— *Leonard Nimoy and Harve Bennett*
In *Star Trek IV: The Voyage Home*, 1986, spoken by Chief Engineer Scott

There is no eel so small but it hopes to become a whale.

— *German proverb*

The species of whale known as the black right whale has four kilos of brains and 1,000 kilos of testicles. If it thinks at all, we know what it is thinking about.

— *Jon Lien, 1995*
Quoted by the Norwegian Telegram Agency
Black right whale is an old term for the northern and southern right whale populations.

Animals

They roam the earth from pole to pole; they are equally at home on a wave-washed coral reef or in an arid desert.

— *Frank Chapman*
In *National Geographic*, June 1913
Speaking about the wide range of habitats that support animal populations.

I collected herds and brought forth their increase. From lands I traveled and hills I traversed, the trees and seeds I noticed and collected.

—*Ashurnasirpal II of Assyria, circa 880 B.C.E.*
Quoted in *Scientific American*
This Mesopotamian king assembled a zoo which included elephants, bears, and dolphins.

Rabbits being of the most profitable creatures in England rightly used.

— *Adolphus Speed, 1659*
Recommending the keeping of rabbits for fur and meat.

A continent overrun with mad rabbits would not be a very cheerful place.
— *Australian government, 1887*
> *Part of an Australian government rejection of Louis Pasteur's proposal to introduce Pasteurella septica (Chicken cholera) to aid in eradicating rabbits. Rabbits had become a serious agricultural nuisance to the continent after their introduction. Starting in 1996, over 100 years later, the Australian government instituted just such a program, spreading Rabbit Calcivirus Disease to control the wild rabbit population.*

All animals are equal, but some animals are more equal than others.
— *George Orwell*
In *Animal Farm*, 1945
> *This is the guiding principle behind keystone species where one species has a far greater influence on the environment than other species. The sentiments of the quote are also a component in the Endangered Species Act.*

Of all the animals, the boy is the most unmanageable.
— *Plato*
In *The Republic*, circa 390 B.C.E.

Do what we can, summer will have its flies. If we walk in the woods, we must feed mosquitoes.
— *Ralph Waldo Emerson*
In *Essays: First Series*, 1841

It is a human characteristic to love little animals, especially if they're attractive in some way.
— *David Gerrold*
In *Star Trek: The Trouble With Tribbles*, 1967
> *Too often, ecological preservation and restoration is decided by the charisma of the species involved.*

O to be a frog, my lads, and live aloof from care.
— *Theoritus*
In *The Reapers*, 300 B.C.E.

It's easy to make a frog jump into water when that's what he wants to do.
— *American proverb*

The ant finds kingdoms in a foot of ground.
— *Stephen Vincent Benet*
In *John Brown's Body*, 1928

Secret, and self-contained, and solitary as an oyster.
> — *Charles Dickens*
> In *A Christmas Carol*, 1843

Cows are my passion.
> — *Charles Dickens*
> In *Dombey and Son*, 1848

South Africa celebrates its livestock on its 10 Rand note, which was published from 1978 until 1990.

Animals are such agreeable friends—they ask no questions, they pass no criticism.
> — *George Eliot*
> *It is left to man to ask the questions and to critique actions.*

I have no doubt that it is a part of the destiny of the human race, in gradual improvement, to leave off eating animals, as surely as the savage tribes have left off eating each other when they come into contact with the more civilized.
> — *Henry David Thoreau*

If God did not intend for us to eat animals, why did he make them taste so good?
> — *Anonymous*

When the bird and the book disagree, always believe the bird.
> — *Birdwatcher's proverb*

In that dawn chorus [of bird songs,] one hears the throb of life itself.
> — *Rachel Carson*
> In *The Sense of Wonder*, 1956

The reptilian idea of fun
Is to bask all day in the sun.
A physiological barrier,
Discovered by Carrier,
Says they can't breathe, if they run.

> — *Richard Cowen*
> In *History of Life*, 1989
> David Carrier reported that the sprawling locomotion of a salamander and lizard forces it to compress each lung alternately as it moves. As a result, it can't breathe if it runs.

Biodiversity

Biodiversity is our most valuable but least appreciated resource.

> — *Edward O. Wilson*
> In *The Diversity of Life*, 1992

How varied are your works, O Lord; in wisdom You have made them all.

> — *Psalms 104:24*

The most wonderful mystery of life may well be the means by which it created so much diversity from so little physical matter.

> — *Edward O. Wilson*
> In *The Diversity of Life*, 1992

It is difficult to believe in the dreadful but quiet war of organic beings going on in the peaceful woods and smiling fields.

> — *Charles Darwin*
> In an 1839 journal entry

I remind you that humans are only a tiny minority in this galaxy.

> — *Max Ehrlich*
> In *Star Trek: The Apple*, 1967
> In fact, humans are only a tiny minority on this planet.

We dwell on a largely unexplored planet.

> — *Edward O. Wilson*
> In *The Diversity of Life*, 1992

I always dreamed that man was a stranger on this planet.
> — *Eric Hoffer*
> In a conversation with Bill Smollen

To plumb the depth of our ignorance, consider that there are millions of insect species still unstudied, most or all of which harbor specialized bacteria.
> — *Edward O. Wilson*
> In *The Diversity of Life*, 1992

The outstanding scientific discovery of the twentieth century is not television, or radio, but rather the complexity of the land organism.
> — *Aldo Leopold*
> In *Conservation*, 1953

The ocean's image as biodiversity's poor cousin has been reassessed lately as a case of marine myopia.
> — *William Stolzenburg*
> In *Nature Conservancy*, May 1999

In communities there are little players and big players, and the biggest players of all are the keystone species. As the name implies, the removal of a keystone species causes a substantial part of the community to change drastically.
> — *Edward O. Wilson*
> In *The Diversity of Life*, 1992

It has become clear that an elite group of species exercises an influence on biological diversity out of all proportion to its numbers.
> — *Edward O. Wilson*
> In *The Diversity of Life*, 1992
> *Referring to keystone species, such as humans, that tend to dominate the environment.*

Ugly bags of mostly water
> — *Robert Sabaroff*
> In *Star Trek, The Next Generation: Home Soil*, 1988
> *A crystal life-form's description of humans.*

Chapter 6

Teaching Science

Science is a way to teach how something gets to be known, what is not known, to what extent things *are* known (for nothing is known absolutely), how to handle doubt and uncertainty, what the rules of evidence are, how to think about things so that judgements can be made, how to distinguish truth from fraud, and from show.

— *Richard P. Feynman*
In a speech titled "The Problem of Teaching Physics in Latin America" printed in *Engineering and Science*, November 1963

Technical Writing

To the devil with those who published before us.

— *Aelius Donatus*
Quoted by St. Jerome, his pupil, circa 300
Originally: Pereant qui ante nos nostra dixerunt.

I will be sufficiently rewarded if when telling it to others you will not claim the discovery as your own, but will say it was mine.

— *Thales, circa 600 B.C.E.*
Quoted in *In Mathematical Circles* by H. Eves

Most of the fundamental ideas of science are essentially simple, and may, as a rule, be expressed in a language comprehensible to everyone.

— *Albert Einstein*
In *The Evolution of Physics*, 1938
The expression of the fundamental ideas might be simple, but the implications rarely are simple.

If you cannot express your ideas in words that I can understand, then perhaps you don't know what you are talking about.

— *Charlie Husson*

You know that I write slowly. This is chiefly because I am never satisfied until I have said as much as possible in a few words, and writing briefly takes far more time than writing at length.

— *Karl Friedrich Gauss*

In science one tries to tell people, in such a way as to be understood by everyone, something that no one ever knew before. But in poetry, it's the exact opposite.

> — *Paul Dirac*
> Quoted in *Mathematical Circles Adieu* by H. Eves

Vigorous writing is concise. A sentence should contain no unnecessary words, a paragraph no unnecessary sentences, for the same reason that a drawing should have no unnecessary lines and a machine no unnecessary parts.

> — *William Strunk*
> In *The Elements of Style,* 1959
> *In other words KISS: Keep It Simple, Stupid.*

I have the conviction that excessive literary production is a social offence.

> — *George Eliot*
> In a letter to Miss Hennell,
> October 28, 1865

Fɪɢ. 629.—Sharpening a pen.

A big book is a big nuisance.

> — *Callimachus*
> In *Fragmenta Incerta,* 260 B.C.E.

From *A Manual of Engineering Drawing for Students and Draftsmen* by Thomas French, 1924

Another damned, thick, square book!

> — *William Henry, Duke of Gloucester, 1781*
> Spoken to the author of *The Decline and Fall of the Roman Empire*

We have a habit in writing articles published in scientific journals to make the work as finished as possible, to cover up all the tracks, to not worry about the blind alleys or describe how you had the wrong idea first, and so on. So there isn't any place to publish, in a dignified manner, what you actually did in order to get to do the work.

> — *Richard Feynman*
> In his *Nobel Lecture,* 1966

Unfortunately what is little recognized is that the most worthwhile scientific books are those in which the author clearly indicates what he does not know; for an author most hurts his readers by concealing difficulties.

> — *Evariste Galois, 1830*
> Quoted in *Mathematical Maxims and Minims* by N. Rose

I do not know what I may appear to the world, but to myself I seem to have been only like a boy, playing on the sea-shore, and diverting myself in now and then finding a smoother pebble or a prettier shell than ordinary, whilst the great ocean of truth lay undiscovered before me.

> — *Sir Isaac Newton*
> Quoted in *Memoirs of Newton* by D. Brewster
> *Except for his papers on optics, every one of Newton's works was published only under pressure from his friends and against his wishes.*

One can measure the importance of a scientific work by the number of earlier publications rendered superfluous by it.

> — *David Hilbert*
> Quoted in *Mathematical Circles Revisited* by H. Eves
> *Unfortunately, too many authors believe that the importance is given by how many earlier publications are contradicted by it.*

It is a good thing for an uneducated man to read books of quotations.

> — *Winston Churchill, 1930*
> *We also think that educated people should read books of quotations.*

It is a great thing to start life with a small number of really good books which are your very own.

> — *Sir Arthur Conan Doyle*
> In *Through the Magic Door*, 1908

I love a broad margin.

> — *Henry David Thoreau*
> In *Walden*, 1854
> *He was referring to his life, but this quote also applies to publishing.*

Punctuation is Power. Use it wisely and for the good of humanity.

> — *Jan Eliot*
> In a cartoon caption in *Stone Soup*, 1999

The obvious is better than obvious avoidance of it.

> — *H. W. Fowler*
> In *Modern English Usage*, 1926

Prefer geniality to grammar.
— *H.W. Fowler and F.G. Fowler*
In *The King's English*, 1906
In other words, when faced with the decision between writing that which is clear and that which is grammatically correct, chose that which is clear.

Put'em down! Facts and Figures!
— *Charles Dickens*
In *The Chimes*, 1844, spoken by Toby "Trotty" Veck

The last thing one knows when writing a book is what to put first.
— *Blaise Pascal*
In *Pensees*, 1670

Writing a textbook can be much like a childhood disease; a single episode is unpleasant, but usually provides lifetime immunity.
— *David J. Weber*

It is very necessary that those who are trying to learn from books the facts of physical science should be enabled by the help of a few illustrative experiments to recognize these facts when they meet them out of doors.
— *James Maxwell*
In the "Introductory Lecture on Experimental Physics"

Man builds no structure which outlives a book.
— *Eugene Fitch Ware*
In "Ironquill" in *The Book*, 1899
Yet, can anyone name a book that has outlived the pyramids of Egypt?

Someone told me that each equation I included in the book would halve the sales.
— *Stephen Hawking*
In *A Brief History of Time*, 1988
One review described Hawking's book as one of the most purchased but least read books of all time.

We have read your manuscript with boundless delight. If we were to publish your paper, it would be impossible to publish a work of lower standard. And as it is unthinkable that in the next thousand years we shall ever see its equal, we are, to our regret, compelled to return your divine composition, and to beg you a thousand times to overlook our short-sight and timidity.

> — *attributed to a modern Chinese economics journal*
> Quoted by Robert Romer in the *American Journal of Physics*

Presentations

Science is voiceless; it is the scientists who talk.

> — *Simone Weil*
> In *On Science, Necessity, and the Love of God*, 1968

It is not that scientists are not listened to; it is that their messages cannot be heard.

> — *Peter Wilcock*
> In *EOS, Transactions, American Geophysical Society*

Speak with the speech of the world, think with the thoughts of the few.

> — *John Hay*
> *The key to excellence is to have novel ideas and to present them clearly.*

How forcible are right words!

> — *Job 6:25*

To be heard, the scientific community must come up with a message that is not only correct, but also simple, direct and coherent. The way to get to this point of clarity—and this cuts to the heart of a scientist's role—is to correctly identify the questions that need to be addressed.

> — *Peter Wilcock*

There is nothing so mysterious as a fact clearly described.

> — *Anonymous*

Get your facts first, and then you can distort them as much as you please.

> — *Mark Twain*

A little inaccuracy sometimes saves tons of explanation.
— *Hector Hugh Munro*
In "The Comments of Moung Ka" in *The Square Egg*, 1924

Although they are only breath, words which I command are immortal.
— *Sappho*

In science, the credit goes to the man who convinces the world, not to the man to whom the idea first occurs.
— *Sir William Osler*

Scientific objectivity must appear to be boring. Scientists are well aware that their work is neither boring nor objective. If it were, very few discoveries would be made. Since the more or less hard edges of scientific inquiry are not involved, social scientists are free to be more categorical about truth, reality and what they call facts. They therefore seek to be more boring than scientists.
— *John Ralston Saul*
In *The Doubter's Companion*, 1994

Students

What students actually *do* is what matters, not what the teachers present... Students are not blank slates.
— *Edward Redish*
In a talk titled "The Role of Physics Education Research in Reforming Undergraduate Education" given at the Physics Revitalization Conference on October 3, 1998
Students need to be actively engaged in their own learning and their misconceptions need to be addressed.

What I am going to tell you about is what we teach our physics students in the third or fourth year of graduate school... It is my task to convince you not to turn away because you don't understand it. You see my physics students don't understand it... That is because I don't understand it. Nobody does.
— *Richard Feynman*
In *QED: The Strange Theory of Light and Matter*, 1990

You never can tell what you have said or done till you have seen it reflected in other people's minds.
— *Robert Frost*
In "Education by Presence" in the *Christian Science Monitor*, December 1925

The business of education is not to make the young perfect in any one of the sciences, but to so open and dispose their minds as may best make them capable of any when they shall apply themselves to it.
— *John Locke*

If we want to understand why our instruction works or doesn't, we have to understand something about how our students' minds function.
— *Edward Redish*
In a lecture titled "Using the Culture of Science to Learn How to Teach Science" given at M.I.T. on May 9, 1999
Advocating the use of research in physics education in order to understand why students consistently fail to understand certain concepts.

There once were four college students taking a class together.

They had done so well on all the quizzes, midterms and labs, etc., that each had an "A" so far for the semester. These four friends were so confident, the weekend before finals, they decided to go out and party with some friends. They had a great time—however, after all the hearty-partying, they slept all day Sunday and didn't make it back to college until early Monday morning. Rather than taking the final then, they decided to find their professor after the final and explain to him why they missed it. They explained that they had gone to a remote mountain cabin for the weekend to study, but, unfortunately, they had a flat tire on the way back, didn't have a spare, and couldn't get help for a long time. As a result, they missed the final. The professor thought it over and then agreed they could make up the final the following day. The guys were elated and relieved. They studied that night and went in the next day at the time the professor had told them. He placed them each in separate rooms, handed each one a test booklet, and told them to begin. They looked at the first problem, worth 5 points. It was something simple.

"Cool," they thought simultaneously, each in his separate room, "this is going to be easy." Each finished the problem and then turned the page.

On the second page was written: (For 95 points): Which tire?
— *Anonymous*

Teachers

In a completely rational society, the best of us would aspire to be teachers and the rest of us would have to settle for something less, because passing civilization along from one generation to the next ought to be the highest honor and the highest responsibility anyone could have.
— *Lee Iacocca*

Nowadays we have too much to teach and too little time to teach it.
—*James Maxwell*
In a lecture at Marischal College, November 3, 1856

A good preparation takes longer than the delivery.
— *E. Kim Nebeuts*

When I had given the same lecture several times I couldn't help feeling that they really ought to know it by now.
—*J.E. Littlewood*
In *A Mathematician's Miscellany*, 1953

If I am given a formula, and I am ignorant of its meaning, it cannot teach me anything, but if I already know it what does the formula teach me?
— *St. Augustine*
In *De Magistro*, circa 400
It is the place of teachers to help impart the meaning of new formulas.

There are two kinds of things that teachers must do well. They can set up environments and situations that are conducive to learning, and they can help students get unstuck. It is difficult to be more specific.
— *Frank Oppenheimer*
In "Teaching and Learning: The AAPT Millikan Lecture" in the *American Journal of Physics*, December 1973

An expert is a man who has made all the mistakes which can be made in a very narrow field.
— *Niels Bohr*

Always listen to experts. They'll tell you what can't be done, and why. Then do it.
— *Robert Heinlein*

An expert is someone who knows some of the worst mistakes that can be made in his subject, and how to avoid them.
— *Werner Heisenberg*
In *Physics and Beyond*, 1971

Expertise in one field does not carry over into other fields. But experts often think so. The narrower their field of knowledge the more likely they are to think so.
— *Robert Heinlein*
In *Time Enough for Love*, 1973

The authority of those who profess to teach is often a positive hindrance to those who desire to learn.
— *Cicero*
In *De Natura Deorum*, circa 45 B.C.E.

Teaching school is but another word for sure and not very slow destruction.
— *Thomas Carlyle*
Quoted in *In Mathematical Circles* by H. Eves

A professor is one who can speak on any subject—for precisely fifty minutes.
— *Norbert Wiener*

Teaching Technique

You cannot teach a man anything; you can only help him to find it for himself.
— *Galileo Galilei*

We are usually convinced more easily by reasons we have found ourselves than by those which have occurred to others.
— *Blaise Pascal*
In *Pensees*, 1670

Ask and learn.
— *Maccabees 6:27*

It is a cliché of our times that what this country needs is an "informed citizenry."… I suggest, rather, that what we need is a *knowledgeable* citizenry. Information, like entertainment, is something someone else provides us… *We cannot be knowledged!* We must all acquire knowledge for ourselves.

> — *Daniel J. Boorstein*
> In remarks at the White House Conference on Library and Information Services, November 1979

How many of those who undertake to educate the young appreciate the necessity of first teaching them how to acquire knowledge?

> — *John Amos Comenius*

Learning is a growth—not a transfer.

> — *Edward Redish*
> In a lecture titled "Using the Culture of Science to Learn How to Teach Science" given at M.I.T. on May 9, 1999

Constant questioning is the first key to wisdom. For through doubt we are led to inquiry, and by inquiry we discern the truth.

> — *Pierre Abelard*
> In *Sic et Non*, 1123

Instruct a lad according to his ways.

> — *Proverbs 22:6*

I am always willing to learn, however I do not always like to be taught.

> — *Winston Churchill*

Bodily exercise, when compulsory, does no harm to the body; but knowledge which is acquired under compulsion obtains no hold on the mind.

> — *Plato*
> In *The Republic*, circa 390 B.C.E.

Should we force science down the throats of those that have no taste for it? Is it our duty to drag them kicking and screaming into the twenty-first century? I am afraid that it is.

> — *George Porter*
> In a speech in September 1986

What may look simple to someone accustomed to a context may be hard to someone new to that context.

— *Edward Redish*

In a talk titled "Building a Science of Teaching Physics: Learning What Works and Why" given at the American Association of Physics Teachers on August 6, 1998

The educational value of…experiments is often inversely proportional to the complexity of the apparatus. The student who uses home-made apparatus, which is always going wrong, often learns more than one who has the use of carefully adjusted instruments, which he is apt to trust, and which he dares not take to pieces.

— *James Maxwell*

In his "Introductory Lecture on Experimental Physics"

The most serious criticism which can be urged against modern laboratory work in Physics is that it often degenerates into a servile following of directions, and thus loses all save a purely manipulative value. Important as is dexterity in the handling and adjustment of apparatus, it can not be too strongly emphasized that it is *grasp of principles*, not *skill in manipulation* which should be the primary object of General Physics courses.

— *Robert A. Millikan, 1903*

Building a good functional understanding requires active intellectual engagement. Hands on activities are not enough. They have to be brains on, as well.

— *Edward Redish*

In a lecture titled "Using the Culture of Science to Learn How to Teach Science" given at M.I.T. on May 9, 1999

Compare this quote with the previous quote by Robert Millikan. The impediments to effective teaching have not changed dramatically in the last century.

The problem with the computer [is] the impedance mismatch between faculty and students… Fancy simulations are not effective when many students lack the concepts needed to interpret them.

— *Edward Redish*

In a talk titled "Using the computer in teaching physics: Can it really help students learn?" given at the American Association of Physics Teachers on January 12, 1999

The worth of a computer in the classroom is only as great as the teacher's ability to use it and the student's ability to understand its results.

Experiments and real world connections first, abstractions later.
— *Edward Redish*
In a talk titled "Building a Science of Teaching Physics: Learning What Works and Why" given at the American Association of Physics Teachers on August 6, 1998

It is nothing short of a miracle that modern methods of instruction have not yet entirely strangled the holy curiosity of inquiry.
— *Albert Einstein*
Quoted in *Return to Mathematical Circles* by H. Eves

Teach to the problems, not to the text.
— *E. Kim Nebeuts*
And teach to the real-world problems, not to the problems on the next test.

Value of Education

The direction in which education starts a man will determine his future life.
— *Plato*
In *The Republic*, circa 390 B.C.E.

Give a man a fish, you feed him for a day; teach him how to fish, you feed him for a lifetime.
— *attributed to Lao-Tzu*
The fundamental truth expressed in this quote has been traced to many different cultures on many different continents.

You live and learn. Or you don't live long.
— *Robert Heinlein*
In *Time Enough for Love*, 1973

I must study politics and war that my sons may have liberty to study mathematics and philosophy. My sons ought to study mathematics and philosophy, geography, natural history, naval architecture, navigation, commerce and agriculture in order to give their children a right to study painting, poetry, music, architecture, statuary, tapestry, and porcelain.
— *John Adams*
In a letter to Abigail Adams, May 12, 1780

Knowledge is of two kinds. We know a subject ourselves, or we know where we can find information upon it.
— *Samuel Johnson, 1775*
> *Most technical classes do not so much teach information as they teach where to find the information. This is why closed-book exams make little sense.*

The truth is found when men are free to pursue it.
— *President Franklin Delano Roosevelt*
In an address at Temple University, February 1936

Live and learn; die and forget it all.
— *Anonymous*

Common sense is in spite of, not as the result of, education.
— *Victor Hugo*

The Way of Inquiry: Much learning does not teach understanding.
— *Heraclitus, circa 500 B.C.E.*
Quoted in *Heraclitus* by Philip Wheelwright

Too much and too little education hinder the mind.
— *Blaise Pascal*
In *Pensees*, 1670

Training is everything. The peach was once a bitter almond; cauliflower is nothing but cabbage with a college education.
— *Mark Twain*

It is, of course, a bit of a drawback that science was invented after I left school.
— *Lord Carrington*
In *The Observer*, January 23, 1983
> *Scientific knowledge will always be moving forward. A good comprehension of the fundamental behavior will enable life-long understanding.*

Education...has produced a vast population able to read but unable to distinguish what is worth reading.
— *George Macaulay Trevelyan*
In *English Social History*, 1942

The more we study, the more we discover our ignorance.
— *American proverb*

Here we tolerate any error so long as truth is left free to combat it.
— *Thomas Jefferson*

One of the benefits of a college education is to show the boy its little avail.
— *Ralph Waldo Emerson*
In *The Conduct of Life*, 1860

Universities

One man's trash is another man's dissertation.
— *David Alsobrook, 1998*
Quoted in the *Boston Globe*
Director of the Bush Presidential Library on the 40 million documents at the library.

A university should be a place of light, of liberty, and of learning.
— *Benjamin Disraeli*
In a speech to the House of Commons, March 11, 1873

A whale ship was my Yale College and my Harvard.
— *Herman Melville*
In *Moby Dick*, 1851

The library serves no purpose unless someone is using it.
— *Jean Lisette Aroeste*
In *Star Trek: All Our Yesterdays*, 1969

University politics are vicious precisely because the stakes are so small.
— *Henry Kissinger, 1973*

Universities hire professors the way some men choose wives—they want the ones the others will admire.
— *Morris Kline*
In *Why the Professor Can't Teach*, 1977

I firmly believe that research should be offset by a certain amount of teaching, if only as a change from the agony of research. The trouble, however, I freely admit, is that in practice you get either no teaching, or else far too much.

> —*J.E. Littlewood*
> In "The Mathematician's Art of Work" in *Littlewood's Miscellany* edited by Bela Bollobas

The Beginning and End of our Didactic will be: To seek and find a method by which the teachers teach less and the learners learn more, by which schools have less noise, obstinacy, and frustrated endeavor, but more leisure, pleasantness, and definite progress.

> —*John Amos Comenius*

The average Ph.D. thesis is nothing but a transference of bones from one graveyard to another.

> —*J. Frank Dobie*
> In *A Texan in England*, 1945
> *Perhaps, but the bones must be transferred in an original manner.*

So far as the mere imparting of information is concerned, no university has had any justification for existence since the popularization of printing in the fifteenth century.

> —*Alfred North Whitehead*
> In *The Aims of Education*

If I were founding a university I would found first a smoking room; then when I had a little more money in hand I would found a dormitory; then after that, or more probably with it, a decent reading room and a library. After that, if I still had more money that I couldn't use, I would hire a professor and get some textbooks.

> —*Stephen Leacock*
> In *Oxford As I See It*
> *The most important aspect of a university is to have discussions and the free-flow of ideas. Classes hold secondary importance.*

Everywhere I go I'm asked if I think the university stifles writers. My opinion is that they don't stifle enough of them.

> —*Flannery O'Connor*

Those involved with higher education tend to confuse growth with life:
they forget that cemeteries grow continuously... Schools are committed
to teaching, not learning, because teaching, unlike learning, can be
industrialized and mechanized; it is easier to control, budget, schedule,
observe, and measure... Teaching is an input to education, not an output,
but our institutions act as though an ounce of teaching is worth at least
a pound of learning.
> — *Russell L. Ackoff*
> In *A Management Scientist Looks at Education and Education
> Looks Back*, 1974

Knowledge, sir, should be free to all!
> — *Stephen Kandel*
> In *Star Trek: I, Mudd*, 1967
> *The character Harry Mudd on the reason he did not pay royalties to the
> owners of patents he had stolen.*

The concept is interesting and well-formed, but in order to earn better
than a 'C', the idea must be feasible.
> — *attributed to a Yale University management professor*
> *In response to student Fred Smith's paper proposing reliable overnight
> delivery service. Smith went on to found Federal Express Corp.*

Chapter 7

The Working Environment

May you live in interesting times.
— *Ancient Chinese Curse*

Committees

Great discoveries and improvements invariably involve the cooperation of many minds.
— *Alexander Graham Bell, 1877*

Interdependence is a higher value than independence.
— *Steven R. Covey*
In *Seven Habits of Highly Effective People*, 1989

A wise man will hear, and will increase learning; and a man of understanding shall attain unto wise counsels.
— *Proverbs 1:5*

If you are going to go looking for evidences of life on other celestial bodies, you need special instruments with delicate sensors for detecting the presence of committees. If there is life there, you will find consortia.
— *Lewis Thomas*
In *The Medusa and the Snail*, 1979

In the course of my life I have submitted a good many ideas to… committees. And I can tell you that their reaction is to see the wrongness, to obliterate 90 per cent of the rightness which the average eye cannot see for the sake of 10 per cent wrongness which the conventional eye always sees. Even though rightness is seen, the new field, the great possibilities that the new idea opens up, will not be visualized.
— *Charles F. Kettering*
In *Fluid Mechanics*, 1962

When it comes to facing up to serious problems, each candidate will pledge to appoint a committee. And what is a committee? A group of the unwilling, picked from the unfit, to do the unnecessary. But it all sounds great in a campaign speech.
—*Richard Long Harkness*

When we ask advice, we are usually looking for an accomplice.
> —*Joseph Louis Lagrange*

Teamwork is essential. It gives the boss more than one target to shoot at.
> — *Craig Fischenich, 1999*

Committee: A cul-de-sac to which ideas are lured and then quietly strangled.
> —*John A. Lincoln*

A committee is a group that keeps minutes and loses hours.
> —*Milton Berle*

We need to listen to one another if we are to make it through this age of apocalypse and avoid the chaos of the crowd.
> — *Chaim Potok*
> Quoted by Steven R. Covey in *Seven Habits of Highly Effective People*

A man who goes alone can start today; but he who travels with another must wait till the other is ready.
> — *Henry David Thoreau*
> In *Walden*, 1854
>
> *The key is to know whether the travel can be done alone or whether a group is required. In addition, travelling with another often makes the travel more enjoyable.*

From each according to his abilities, to each according to his needs.
> — *Karl Marx*
> In *Critique of the Gotha Program*, 1875
>
> *Although applied to communist economic doctrine, the philosophy is also applicable to working groups.*

Organization doesn't really accomplish anything. Plans don't accomplish anything, either. Theories of management don't much matter. Endeavors succeed or fail because of the people involved. Only by attracting the best people will you accomplish great deeds.
> — *Colin Powell*
> In a lecture titled "Great Lessons in Leadership," 1998

Science has now been for a long time—and to an ever-increasing extent—a collective enterprise. Actually, new results are always, in fact, the work of specific individuals; but, save perhaps for rare exceptions, the value of any result depends on such a complex set of interrelations with past discoveries and possible future researches that even the mind of the inventor cannot embrace the whole.

> — *Simone Weil*
> In *Oppression and Liberty*, 1955

Planning and Implementation

We trained hard, but it seemed that every time we were beginning to form up into teams, we would be reorganized. I was to learn later in life that we tend to meet any new situation by reorganizing; and a wonderful method it can be for creating the illusion of progress while producing confusion, inefficiency and demoralization.

> — *Petronius Arbiter, 210 B.C.E.*
> Some committees have changed little in the last two millennia.

Rome did not create a great empire by having meetings, they did it by killing all those who opposed them.

> — *Anonymous*

Victory… will depend on execution not plans.

> — *George Patton*
> In a letter to Dwight Eisenhower, 1926. Quoted in *Citizen Soldier* by Stephen Ambrose.
> While Patton was referring to warfare, the invocation to action is appropriate for other endeavors as well. (And yes, the date is correct.)

Plans are nothing; planning is everything.

> — *Dwight Eisenhower*

The success of any operation is as much dependent on execution as it is on planning and concept.

> — *Anonymous*

You need a plan for everything, whether it's building a cathedral or a chicken coop.
— *John Goddard*

And they planned and Allah also planned, and Allah is the best of planners.
— *The Koran 3:54*

And they have planned a very great plan.
— *The Koran 71:22*

Chance favors only the prepared mind.
— *Louis Pasteur*

Insufficient facts always invite danger.
— *Gene L. Coon and Carey Wilbur*
In *Star Trek: Space Seed*, 1967

No man is justified in doing evil on the ground of expediency.
— *President Theodore Roosevelt*
In *The Strenuous Life*, 1900

Delay is preferable to error.
— *Thomas Jefferson*
In a letter to George Washington, May 16, 1792
The balance between schedule and product generates continued debate.

Many survival case histories show that stubborn, strong willpower can conquer many obstacles.
— *U.S. Army Survival Manual, FM 21-76, 1998*

The greatest enemies in a combat survival situation are fear and panic. If uncontrolled, they can destroy your ability to make an intelligent decision.
— *U.S. Army Survival Manual, FM 21-76, 1998*

Always do right—this will gratify some and astonish the rest.
— *Mark Twain*
In a message to the Young People's Society, New York City on February 16, 1901.
President Harry S. Truman had this remark framed behind his desk in the Oval Office.

It takes considerable knowledge just to realize the extent of your
own ignorance.
> *— Thomas Sowell, 1999*
> Quoted in *Readers Digest*

In every work of genius we recognize our own rejected thoughts;
they come back to us with a certain alienated majesty.
> *— Ralph Waldo Emerson*

We can lick gravity, but sometimes the paperwork is overwhelming.
> *— Wernher von Braun*

If you don't much care where you want to get to, then it doesn't matter
which way you go.
> *— Lewis Carroll*
> In *Alice in Wonderland*, the Cheshire Cat, 1865

In science, error is still recognized as a permanent characteristic of
progress... Errors that lead to altered policies are the building block
of civilization.
> *—John Ralston Saul*
> In *The Doubter's Companion*, 1994

Seek simplicity, and distrust it.
> *— Alfred North Whitehead*

The most dangerous thing in the world is to try to leap a chasm in
two jumps.
> *— William Lloyd George*

Hofstadter's Law: It always takes longer than you expect, even when you
take into account Hofstadter's Law.
> *— Douglas Hofstadter*
> In *Godel, Escher, Bach*, 1979

Cheops' Law: Nothing *ever* gets built on schedule or within budget.
> *— Robert Heinlein*
> In *Time Enough for Love*, 1973
> *Please note, however, that the Panama Canal was finished ahead of schedule
> and under budget.*

There were these two managers who decided they would go moose hunting. As it happened, they shot a moose. They were about to drag the animal by the hind legs when a biologist and an engineer came along.

The Biologist said, "You know, the hair follicles on a moose have a grain to them that causes the hair to lie toward the back."

The Engineer said, "The way you are dragging that moose, it increases your coefficient of friction by a tremendous amount. If you pull from the other end, you will find the work required to be quite minimal."

The managers thanked the two and started dragging the moose by the antlers. After about an hour, one manager said, "I can't believe how easy it is to move this moose this way. I sure am glad we ran across those two."

"Yeah," said the other. "But we're getting further and further away from our truck."

> — *Anonymous*
>> *Moral: It is not enough just to have correct data, you must be able to apply it.*

Human Behavior

What a piece of work is man.
> — *William Shakespeare*
> In *Hamlet*, 1601

I know of no more encouraging fact than the unquestionable ability of man to elevate his life by conscious endeavor.
> — *Henry David Thoreau*
> *One should endeavor to restore that which man has damaged.*

Being responsible sometimes means pissing people off.
> — *Colin Powell*
> In a talk titled "Great Lessons in Leadership," 1998

In all fields of human endeavor, 80 percent of the results flow from 20 percent of the activities.
> — *Vilfredo Pareto*
> Quoted by Steven R. Covey in *Seven Habits of Highly Effective People*

The difference between a good man and a bad is the choice of cause.
> — *William James*

A trapped animal makes the most noise.
> —*American proverb*

No one can make you feel inferior without your consent.
> —*Eleanor Roosevelt*

If you gave people a choice, there would be chaos.
> —*Mac McDonald*
> Quoted in *McDonald's—Behind the Arches* by John F. Love.

Convictions are more dangerous foes of truth than lies.
> —*Friedrich Nietzsche*

There are three types of people in the world: those who have wishbones, those who have funny bones, and those who have backbones.
> —*American proverb*

In critical moments men sometimes see exactly what they wish to see.
> —*Judy Burns and Chet Richards*
> In *Star Trek: The Tholian Web*, 1968

The corporate world.

Such is the human race. Often it does seem such a pity that Noah and his party did not miss the boat.
> —*Mark Twain*
> In *What Is Man?*, 1907
> *Not all human behavior is admirable.*

One man cannot do right in one department of life whilst he is occupied doing wrong in any other department. Life is one indivisible whole.
> —*Mohandas Gandhi*

Because they could see it almost any night perhaps they will never see it.
> *— Rachel Carson*
> In *The Sense of Wonder*, 1956
> *Speaking about the stars.*

If we were not provided with the knack of being wrong, we could never get anything useful done.
> *— Lewis Thomas*
> In *The Medusa and the Snail*, 1979
> *In other words, people tend to learn best through "trial and error."*

The first human being who hurled an insult instead of a stone was the founder of civilization.
> *— Attributed to Sigmund Freud*

A reasonable amount of fleas is good fer a dog—keeps him from broodin' on bein' a dog.
> *— Edward Noyes Westcott*
> In *David Harum*, 1898

The man who is a pessimist before 48 knows too much; if he is an optimist after it, he knows too little.
> *— Mark Twain*
> In his notebook entry for December 1902

A human being should be able to change a diaper, plan an invasion, butcher a hog, conn a ship, design a building, write a sonnet, balance accounts, build a wall, set a bone, comfort the dying, take orders, give orders, cooperate, act alone, solve equations, analyze a new problem, pitch manure, program a computer, cook a tasty meal, fight efficiently, die gallantly.
> *— Robert Heinlein*
> In *Time Enough for Love*, 1973

Man is full of desires: he loves only those who can satisfy them all. "This man is a good mathematician," someone will say. But I have no concern for mathematics; he would take me for a proposition. "That one is a good soldier." He would take me for a besieged town. I need, that is to say, a decent man who can accommodate himself to all my desires in a general sort of way.
> *— Blaise Pascal*
> Quoted in *The Viking Book of Aphorisms* by W. Auden and L. Kronenberger

Have fun in your command. Don't always run at a breakneck pace. Take leave when you've earned it: Spend time with your families. Surround yourself with people who take their work seriously, but not themselves, those who work hard and play hard.

> — *Colin Powell*
> In a lecture titled "Great Lessons in Leadership," 1998

The only person who never makes a mistake is the person who does nothing.

> — *Charlie Husson*

There comes a time in every rightly constructed boy's life when he has a raging desire to go somewhere and dig for hidden treasure.

> — *Mark Twain*
> In *Tom Sawyer*, 1876

The more I see of men, the better I like my dog.

> — *Blaise Pascal*

Communication

If we had more time for discussion we should probably have made a great many more mistakes.

> — *Leon Trotsky*

Strong words are required for weak principles.

> — *Doug Horton*

The one who listens does the most work, not the one who speaks.

> — *Steven R. Covey*
> In *Seven Habits of Highly Effective People*, 1989

People always get what they ask for; the only trouble is that they never know, until they get it, what it actually is that they have asked for.

> — *Aldous Huxley*

The Moving Finger writes; and, having writ,
Moves on: nor all thy Piety nor Wit
   Shall lure it back to cancel half a Line,
Nor all thy Tears wash out a Word of it.
     — Omar Khayyám
       In *The* Rubáiyyát, circa 1100, translated by Edward FitzGerald

Criticism comes easier than craftsmanship.
     — Zeuxis, 400 B.C.E.

It is not necessary to understand things in order to argue about them.
     — Pierre Augustin Caron de Beaumarchais

Sir, I have found you an argument. I am not obliged to find you
an understanding.
     — Samuel Johnson
       Quoted in *The Life of Samuel Johnson* by J. Boswell, 1784

Guard your tongue from speaking evil, and your lips from deceitful speech.
     — Psalms 34:14

What's on a sober man's mind is on a drunk man's tongue.
     — Russian proverb

I can retract that I did not say, but I cannot retract that I already have said.
     — Solomon Ibn Gabirol
       In *Pearls of Wisdom*

Tact consists in knowing how far to go in going too far.
     —Jean Cocteau

Whoever in discussion adduces authority uses not intellect but memory.
     — Leonardo da Vinci
       In his notebooks

Speak softly, and carry a big stick; you will go far.
     —Theodore Roosevelt

[Napoleon] directed Bourrienne to leave all his letters unopened for three weeks, and then observed with satisfaction how large a part of the correspondence had thus disposed of itself, and no longer required an answer.

> — *Ralph Waldo Emerson*
> In "Napoleon" in *Representative Men,* 1850

Communication across the revolutionary divide is inevitably partial.

> — *Thomas S. Kuhn*

Let your speech be always with grace, seasoned with salt.

> — *The Epistle of Paul to the Colossians 4:6*

A word fitly spoken is like apples of gold in a setting of silver.

> — *Proverbs 25:11*

We change one communication for another communication.

> — *The Koran 16:101*
> *No matter what form of communication we use, we must always communicate.*

The real art of conversation is not only to say the right thing at the right place but to leave unsaid the wrong thing at the tempting moment.

> — *Dorothy Nevill*

What we've got here is a failure to communicate.

> — *Donn Pearce*
> In the movie *Cool Hand Luke,* 1967

Today, if you are not confused, you are just not thinking clearly.

> — *Irene Peter*

Cynicism: An effective social mechanism for preventing communication.

> — *John Ralston Saul*
> In *The Doubter's Companion,* 1994

I was gratified to be able to answer promptly, and I did. I said I didn't know.

> — *Mark Twain*
> In a letter to William Dean Howells, December 1877

The day soldiers stop bringing you their problems is the day you have stopped leading them. They have either lost confidence that you can help them or concluded that you do not care. Either case is a failure of leadership.

> — *Colin Powell*
> In a lecture titled "Great Lessons in Leadership," 1998

A man is flying in a hot air balloon and realizes he is lost. He reduces height, spots a man down below and asks, "Excuse me, can you help me? I promised to return the balloon to its owner, but I don't know where I am."

The man below says: "You are in a hot air balloon, hovering approximately 350 feet above mean sea level and 30 feet above this field. You are between 40 and 42 degrees north latitude, and between 58 and 60 degrees west longitude."

"You must be an engineer," says the balloonist.

"I am," replies the man. "How did you know?"

"Well," says the balloonist, "everything you have told me is technically correct, but I have no idea what to make of your information, and the fact is I am still lost."

The man below says, "You must be a manager."

"I am," replies the balloonist, "but how did you know?"

"Well," says the engineer, "you don't know where you are, or where you are going. You have made a promise which you have no idea how to keep, and you expect me to solve your problem. The fact is you are in the exact same position you were in before we met, but now it is somehow my fault."

> — *Anonymous*
> *Moral 1: Not only do you have to know how to answer a question, but also you have to be able to ask it.*
> *Moral 2: Communication is not just the exchange of ideas, it is the exchange of understanding.*

Call to Action

Go, sir, gallop, and don't forget that the world was made in six days.
You can ask me for anything you like, except time.
> — *Napoleon Bonaparte, 1803*

There are a thousand hacking at the branches of evil to one striking at
the root.
> — *Henry David Thoreau*
> In *Walden*, 1854

All that you find within your ability to do, act upon it.
> — *attributed to King Solomon*
> In *Ecclesiastes* 9:10

It is not enough to have a good mind. The main thing is to use it well.
> — *Rene Descartes*
> In *Discours de la Methode*, 1637

Be the change that you want to see in the world.
> — *Mohandas Gandhi*

Even if you're on the right track, you'll get run over if you just sit there.
> — *Will Rogers*

Look at a day when you are supremely satisfied at the end. It's not a day
when you lounge around doing nothing. It's when you've had everything to do and you've done it.
> — *Margaret Thatcher*

If you can fill the unforgiving minute
With sixty seconds worth of distance run
Yours will be the Earth and everything that's in it.
> — *Rudyard Kipling*
> In *If*, 1910

What we know is not much. What we do not know is immense.
> — *Pierre-Simon de Laplace*
> Quoted in the *Budget of Paradoxes*
> *These are attributed as being Laplace's last words, but there are other phrases also attributed to being Laplace's last words.*

Vengeance is thine O God, but make me thy weapon of destruction.
> — *Archie Fripp, Sr., 1944*

Iron rusts from disuse, stagnant water loses its purity, and in cold weather becomes frozen, even so does inaction sap the vigors of the mind.
> — *Leonardo da Vinci*

Never mistake motion for action.
> — *Ernest Hemingway*

Well done is quickly done.
> — *Augustus Caesar*
> In *Suetonius, Augustus*, sec. 25

My motto is "Ready, fire, aim," and you give me lectures and make sly fun of me about that. But yours is "Ready, aim, aim, aim, aim, aim, aim, aim..."
> — *Tom Wolfe*
> In *Man In Full*, 1998

The fruit of labors, in lives to come, Is threefold for all men,—Desirable, and Undesirable and mixed of both; But no fruit is at all where no work was.
> — In the *Bhagavadgita*, circa 100 C.E., spoken by Krishna

He who desires but acts not, breeds pestilence.
> — *Henry Blake*

"If it ain't broke, don't fix it" is the slogan of the complacent, the arrogant or the scared. It's an excuse for inaction, a call to non-arms.
> — *Colin Powell*
> In a lecture titled "Great Lessons in Leadership," 1998

Kill or be killed, eat or be eaten, was the law.
>—*Jack London*
>>In *Call of the Wild*, 1903

A ship in harbor is safe, but that is not what ships are built for.
>—*John A. Shedd*

To every thing there is a season, and a time to every purpose under the heaven.
>—*Ecclesiastes 3:1*

Life has two rules: Number 1, Never Quit! Number 2, always remember rule number 1.
>—*Duke Ellington*

Do not be afraid of opposition; remember, a kite rises against the wind.
>—*American proverb*

A still river never finds the ocean.
>—*American proverb*

A running river never freezes.
>—*American proverb*

Calm seas do not make good sailors.
>—*American proverb*

There is no substitute for hard work.
>—*Thomas Edison*

I find that the harder I work, the more luck I seem to have.
>—*Thomas Jefferson*

Necessity is the mother of invention.
>—*Plato*
>>In *The Republic*, circa 390 B.C.E.

"Necessity is the mother of invention" is a silly proverb. "Necessity is the mother of futile dodges" is much nearer the truth.
> — *Alfred North Whitehead*

Furious activity is no substitute for understanding.
> — *H. H. Williams*

One fire burns out another's burning.
> — *William Shakespeare*
> In *Romeo and Juliet*, 1595

Press on: Nothing in the world can take the place of perseverance. Talent will not; nothing is more common than unsuccessful men with talent. Genius will not; unrewarded genius is almost a proverb. Education will not; the world is full of educated derelicts. Persistence and determination alone are omnipotent.
> — *Calvin Coolidge*

You don't know what you can get away with until you try.
> — *Colin Powell*
> In a lecture titled "Great Lessons in Leadership," 1998

What does not destroy me, makes me stronger.
> — *Friedrich Nietzsche*

Nobody ever forgets where he buried the hatchet.
> — *Kin Hubbard*

You don't invent your mission, you detect it.
> — *Victor Frankl*
> In *Man's Search for Meaning*

40 seconds? But I want it NOW!
> — *Matt Groening*
> In *The Simpsons*
> *Homer Simpson responding to microwave instructions.*

Q: What did the snail say when he climbed on the back of the turtle and took a ride?

A: Weeeeeeeee!

Moral: It is the relative feeling of progress that most people look for.

Trust

Forgive your enemies, but never forget their names.

— *President John F. Kennedy*

Trust in Allah, but tie your camel.

— *Arabian proverb*

He who mistrusts most should be trusted least.

— *Theognis*

Make friendship with the wolf, but keep your ax ready.

— *Russian proverb*

Beware of false prophets, which come to you in sheep's clothing, but inwardly they are ravening wolves.

— *Matthew 7:15*

The enemy is anybody who's going to get you killed, no matter which side he's on.

— *Joseph Heller*

In God we trust—everyone else must provide documentation.

— *Anonymous*

No counsel is more trustworthy than that which is given upon ships that are in peril.

— *Leonardo da Vinci*
In his notebooks

Pretension

The higher the monkey climbs, the more he shows his ass.
— *Thomas Watson Jr.*
    President of IBM and ambassador to Russia.

A man is like a fraction whose numerator is what he is and whose denominator is what he thinks of himself. The larger the denominator the smaller the fraction.
— *Leo Tolstoy*

A very popular error—having the courage of one's convictions: rather it is a matter of having the courage for an attack upon one's convictions.
— *Friedrich Nietzsche*

People who know little talk much; people who know much talk little.
— *American proverb*

People who wouldn't think of talking with their mouths full often speak with their heads empty.
— *American proverb*

An ounce of pretension is worth a pound of manure.
— *Anonymous*

The ambitious bullfrog puffed and puffed until he burst.
— *American proverb*

It is with narrow-souled people as with narrow-necked bottles; the less they have in them, the more noise they make in pouring out.
— *American proverb*

Many a person thinks he is hard-boiled when he is only half-baked.
— *American proverb*

Never let your ego get so close to your position that when your position goes, your ego goes with it.

— *Colin Powell*
In a lecture titled "Great Lessons in Leadership," 1998

If I were a medical man, I should prescribe a holiday to any patient who considered his work important.

— *Bertrand Russell*
In *The Autobiography of Bertrand Russell, 1967*

It was the final examination for a large class. The professor was very strict and told the class that any exam that was not on his desk in exactly two hours would not be accepted and the student would fail. A half hour into the exam, a student rushed in and asked the professor for an exam booklet.

"You're not going to have time to finish this," the professor said, as he handed the student a booklet.

"Yes I will," replied the student. He then took a seat and began writing.

After two hours, the professor called for the exams, and the students filed up and handed them in. All except for the late student, who continued writing. Half an hour later, the last student came up to the front of the lecture hall where the professor was sitting behind the desk, casually reading a book with his feet up on a stool. He attempted to put his exam on the stack of exam booklets already there.

"No you don't. I'm not going to accept that. It's late," the professor said, turning the page in his book.

The student looked incredulous and angry. "Do you know who I am?"

"Nooooo, as a matter of fact I don't," replied the professor with an air of sarcasm in his voice.

"Do you KNOW who I AM?" the student asked again, poking his own chest with his finger as he leaned intimidatingly over the table.

"No, and I don't care." replied the professor with an air of superiority.

The student quickly lifted the stack of completed exams and stuffed his in the middle.

"Good!" he said, and walked out of the room.

Money and Greed

Competition brings out the best in products and the worst in men.
— *David Sarnoff*
Founder of NBC and credited with introducing commercial TV broadcasting.

Behind every great fortune there is a crime.
— *Honoré de Balzac*

I cannot afford to waste my time making money.
— *Jean Louis Agassiz*

The battle of competition is fought by cheapening of commodities.
— *Karl Marx*
In *Capital*, 1867–1883

They are the kind of people who don't care whether science is used for production or devastation so long as they grow rich themselves.
— *Leslie Greener*
In *Moon Ahead*, 1951
Description of the "bad guys" in this children's book. Also a description of some of the "bad guys" in real life.

Enough… is never enough.
— *Ira Steven Behr*
In *Ferengi Rule of Acquisition* number 97 from *Star Trek: Deep Space 9*, 1993
The Ferengi are a race of aliens whose entire civilization is centered upon maximizing profit.

Even in the worst of times, someone turns a profit.
— *Ira Steven Behr*
In *Ferengi Rule of Acquisition* number 162 from *Star Trek: Deep Space 9*, 1993

A wealthy man can afford anything except a conscience.
— *Ira Steven Behr*
In *Ferengi Rule of Acquisition* number 261 from *Star Trek: Deep Space 9*, 1993

The sacrifice Which Knowledge pays is better than gifts Offered by wealth.
> — In the *Bhagavadgita*, circa 100 C.E., spoken by Krishna

If the camel gets his nose into the tent, his body will soon follow.
> — *Arab proverb*

Buy when there is blood in the streets.
> — *Stock market proverb*

You can be young without money, but you can't be old without it.
> — *Tennessee Williams*

Greed…is good. Greed is right. Greed works.
> — *Gordon Gekko*

A hasty angler loses the fish.
> — *American proverb*

The bitterness of poor quality remains long after the sweetness of meeting the schedule is forgotten.
> — *Anonymous*

One who loves money will never be satisfied with money.
> — *attributed to King Solomon*
> In Ecclesiastes 5:10

Wealth consists not in having great possessions, but in having few wants.
> — *Epicurus*

Good friends, good books and a sleepy conscience: this is the ideal life. (The conviction of the rich that the poor are happier is no more foolish than the conviction of the poor that the rich are.)
> — *Mark Twain*

The codfish lays ten thousand eggs,
　The homely hen lays one.
The codfish never cackles
　To tell you what she's done.
And so we scorn the codfish,
　While the humble hen we prize,
Which only goes to show you
　That it pays to advertise.
　　　　　— In It Pays to Advertise

Common Sense

There are 40 kinds of lunacy, but only one kind of common sense.
　　　　— West African proverb
　　　　Although some days there seems to be far more than just 40 kinds of lunacy.

It ain't what you know, it's what you know that ain't so that causes trouble.
　　　　— Edward Redish
　　　　In a talk titled "Making Sense of What Happens in Physics Classes: Analyzing
　　　　Student Learning" given at the American Physical Society on March 24, 1999

Go placidly amid the noise and haste and remember what peace there
may be in silence.
　　　　— Plaque in St. Paul's Church, Baltimore, MD. Founded 1692

It's such a fine line between stupid and clever.
　　　　— Michael McKean, Christopher Guest, Rob Reiner, and Harry Shearer
　　　　In *This Is Spinal Tap*, 1984

Do not dress in leaf-made clothes when going to put out a fire.
　　　　— Chinese proverb

Odds and Ends

Miscellaneous is always the largest category.
> —*Joel Rosenberg*
> In *The Warrior Lives*

It's a sad day for American capitalism, when a man can't fly a midget on a kite over Central Park.
> —*Jim Moran, 1998*
> Quoted in *Time* Magazine
> *Spoken after being stopped by the New York Police Department from flying "uniformed midgets" from large kites over Central Park as part of an advertising campaign.*

Let us endeavor to live that when we come to die even the undertaker will be sorry.
> —*Mark Twain*

Men who are unhappy, like men who sleep badly, are always proud of the fact.
> —*Bertrand Russell*

Maybe, just once, someone will call me 'sir' without adding, 'you're making a scene'.
> —*Matt Groening*
> In *The Simpsons*, Homer Simpson

No, no, no Lisa. If adults don't like their jobs, they don't go on strike. They just go in everyday and do it really half-assed.
> —*Matt Groening*
> In *The Simpsons*, Homer Simpson

Life is what happens to you while you're busy making other plans.
> —*John Lennon*

We have committed the Golden Rule to memory; let us commit it to life.
> —*Edwin Markham*

_____ are like snowmen, fun at first but not much future.
— *Anonymous*
This is a fill-in-the-blank quotation.

Back of every noble life there are principles that have fashioned it.
— *George H. Lorimer*

It is impossible for us to break the law. We can only break ourselves against the law.
— *Cecil B. De Mille*

If at first you don't succeed, well, so much for sky diving.
— *Victor Reilly*

If men can run the world, why can't they stop wearing neckties? How intelligent is it to start the day by tying a little noose around your neck?
— *Linda Ellerbee*

Stress: When you wake up screaming and you realize you haven't fallen asleep yet.
— *Anonymous*

References

The quotes in this book were found in a variety of sources. The following is a list of the more useful references and they should be your starting points if you are seeking to find new quotes.

21st Century Dictionary of Quotations. Edited by the Princeton Language Institute, Dell Publishing, 1993.

Bartlett, John, *Familiar Quotations.* Little, Brown and Company, Boston, Massachusetts, 1955.

Beck, A. M., Bortz, A., Lynch, C. W., Mayo, L, and Weld, R. L., *The Corps of Engineers: The War Against Germany.* Center of Military History. United States Army, 1988.

Beckmann, Petr, *A History of Pi.* St. Martin's Press, New York, 1971.

Beebe, Lucius, *Highball, A Pageant of Trains.* Bonanza Books, New York, 1945.

Biedenharn, D. S., Elliott, C. M., Watson, C. C., *WES Stream Investigation and Streambank Stabilization Handbook.* United States Army Corps of Engineers, 1997.

Bras, R. L., *Hydrology, an Introduction to Hydrologic Science.* Addison-Wesley Publishing Company, 1990.

Burn, Barbara, *North American Wildflowers.* The National Audubon Society Collection Nature Series, Gramercy Books, 1984.

Carson, Rachel, *The Sense of Wonder.* The Nature Company, Berkeley, California, 1990.

Chow, Ven Te, Maidment, David, and Mays, Larry, *Applied Hydrology.* McGraw-Hill Publishing, 1988.

Cooper, C. David and Alley, F. C. *Air Pollution Control—A Design Approach.*, Waveland Press, 1986.

Cortright, E.M., *Exploring Space with a Camera.* NASA SP-168, Office for Technical Information Division, Washington D.C., 1968.

Covey, Steven R., *Seven Habits of Highly Effective People.* Simon and Schuster, 1989.

De Camp, L. Sprague, *The Ancient Engineers.* Dorset Press, New York, 1990.

The Flood Control Challenge: Past, Present, and Future; Proceedings of a National Symposium. New Orleans, Louisiana, September 26, 1986, Public Works Historical Society, 1986.

The Floods of March 1936 in Pennsylvania. Prepared by the Commonwealth of Pennsylvania Department of Forests and Waters in cooperation with the United States Geological Survey, 1936.

Fowles, Barry With, *Builders and Fighters, U.S. Army Engineers in World War II.* Office of History, U.S. Army Corps of Engineers, Fort Belvoir, Virginia, 1992.

Frankl, Victor, *Man's Search for Meaning: An Introduction to Logotherapy*. Simon & Schuster Trade Books, New York, 1984.

French, R. H., *Open Channel Hydraulics*. McGraw-Hill Book Company, 1985.

French, Thomas E. *A Manual of Engineering Drawing for Students and Draftsmen*. McGraw-Hill Book Company, New York, 1924.

From Sea to Shining Sea. The Presidents Council on Recreation and Natural Beauty, U.S. Government Printing Office, 1968.

Gleason, Norma, *Proverbs from Around the World*. Carol Publishing Group, 1992.

Gleick, James, *Chaos: Making a New Science*. Penguin Books, 1987.

Gore, Al, *Earth in the Balance—Ecology and the Human Spirit*. Houghton Mifflin Company, 1992.

Greener, Leslie, *Moon Ahead*. The Viking Press, New York, 1951.

Groves, Major General Richard H. "The Agnes Disaster," *The Military Engineer*, January–February, 1973.

Hayden, M., *The Book of Bridges*. Galahad Books, New York, 1976.

Hendrickson, Robert, *Salty Words*. Hearst Marine Books, New York, 1984.

Verhagen, Joachim at http://www.xs4all.nl/~jcdverha/scijokes/index.html

Kadlec, R. H. and Knight, R. L., *Treatment Wetlands*. CRC Lewis Publishers, 1996.

Leopold, Aldo, *A Sand County Almanac*. Oxford University Press, Inc., 1949.

Maggio, Rosalie, *The Beacon Book of Quotations by Women*. Beacon Press, Boston, Massachusetts, 1994.

Moncur, Michael at http://www.starlingtech.com/quotes/

Mieder, W., Kingsbury, S. A., and Harder, K. B. *A Dictionary of American proverbs*. Oxford University Press, 1992.

Millikan, Robert A. and Gale, Henry G. *A First Course in Physics.*, Ginn & Company, Boston, 1906.

Poff, N. L., Allen, J. D., Bain, M. B., Karr, J. R., Prestegaard, K. L., Richter, B. D., Sparks, R. E., and Stromberg, J. C. "The Natural Flow Regime," *Bioscience* 47(11), 1997.

Rabeni, Charles F. and Sowa, Scott P. "Integrating biological realism into habitat restoration and conservation strategies for small streams," *Canadian Journal of Fisheries and Aquatic Science*, 53, 1996.

Redish, Joe at www2.physics.umd.edu/~redish/Money/.

Reisner, Marc, *Cadillac Desert—The American West and its Disappearing Water*. Penguin Books, 1986.

Reuss, M., *Designing the Bayous, 1900–1995*. Office of History, U.S. Army Corps of Engineers, 1998.

Rosen, H. and Reuss, M., *The Flood Control Challenge: Past, Present, and Future*. Public Works Historical Society, 1988.

Rosgen, Dave and Silvey, H. Lee, *Applied River Morphology*. Wildland Hydrology, North Logan, Utah, 1996.

Sackett, Susan, Goldstein, Fred, and Goldstein, Stan, *Star Trek Speaks*. Pocket Books, New York, 1979.

Samuels, Amy and Tyack, Peter "Flukeprints: A history of studying cetacean societies.," In *Cetacean societies: field studies of whales and dolphins*, edited by J. Mann, R. Connor, P.L. Tyack, and H. Whitehead. University of Chicago Press, Chicago, Illinois, in press.

Saul, John Ralston, *The Doubter's Companion*. Viking Penguin, New York, 1994.

Schoklitsch, Armin, *Hydraulic Structures—A Text and Handbook*. Translated by Samuel Shulits, Published by the American Society of Mechanical Engineers, 1937.

Sherrin, Ned, *The Oxford Dictionary of Humorous Quotations*. Oxford University Press, 1995.

Stone, Robert, *GreenPrints—The Weeders Digest*. P.O. Box 1355, Fairview, North Carolina, Published quarterly.

Teller, Walter Magnes. *On the River—A Variety of Canoe and Small Boat Voyages*. Sheridan House, Inc, Dobbs Ferry, New York, 1998.

Thomas, David, *Archaeology*. Holt, Rinehard and Winston, Inc., Chicago, 1989.

Thomas, Lewis, *The Medusa and the Snail*. Viking Press, New York, 1979.

Tropical Storm Agnes. United States Army Corps of Engineers publication, 1972.

U.S. Army Survival Manual, FM 21-76. Department of the Army, Dorset Press, New York, 1998.

Wilson, Edward O., *In Search of Nature*. Island Press/Shearwater Books, Washington, D.C., 1996.

Wilson, Edward O., *The Diversity of Life*. Belknap Press of Harvard University Press, Cambridge, Massachusetts, 1992.

Biographies

By their very nature, quotes are taken out of context. The following very brief biography of many of the speakers featured in this book should help provide some context to their quotes.

Edward Abbey	(1927–1989), American author.
Donald P. Abbott	Marine biologist and author.
Pierre Abelard	(1079–1142?), French philosopher and theologian.
Ackoff	(1919–), American developer of management theories.
John Adams	(1735–1826), second president of the United States.
Alfred Adler	(1870–1937), Austrian psychologist and psychiatrist.
Jean Louis Agassiz	(1807–1873), Swiss-American naturalist and professor who introduced the idea of Ice Ages.
J. Alcock	American behavioral biologist and author.
Tim Allen	(1953–), American comedian and actor.
David Alsobrook	Director of the Bush Presidential Library.
Jeanne Altmann	American behavioral biologist best known for defining the proper methods for sampling behavior.
Petronius Arbiter	(–66 C.E.), Roman author who was considered a good judge of elegance.
Archimedes	(287–212 B.C.E.), Greek mathematician and inventor who wrote important works on mathematics and mechanics.
Aristotle	(384–322 B.C.E.), Greek philosopher who shaped many of the early notions of nature.
Neil Armstrong	(1930–), First human to step onto the moon.
Ashurnasirpal II	(884–859 B.C.E.), King of Assyria.
Isaac Asimov	(1920–1992), Russian-born American science-fiction author and biochemistry professor.
W.H. Auden	(1907–1973), British-American poet and playwright.
Saint Augustine	(354–430), Catholic Bishop and author.
Augustus Caesar	(63 B.C.E.–14 C.E.), First emperor of Rome.
Marcus Aurelius	(121–80 B.C.E.), Roman emperor and philosopher.
Charles Babbage	(1792–1871), British mathematician and inventor who designed and built mechanical computing machines.
Babylonian Talmud	Jewish commentary on the Hebrew Bible (Torah), written 300–600 C.E..
Roger Bacon	(1214?–1294), English philosopher and scientist.
Walter Bagehot	(1826–1877), British economist and journalist.
Parakrama Bahu	(1100s), King of Sri Lanka.
Liberty Hyde Bailey	(1858–1954), American professor of botany who studied genetics, plant pathology, and modern agriculture.
Nancy Banks-Smith	Contemporary British columnist.
James Barnerd	(–1768), sailor.
Patrick Bateson	American behavioral biologist.

Spencer Beebe	American conservationalist and president of Ecotrust.
Steven Behr	Contemporary science-fiction author.
Alexander Graham Bell	(1847–1922), American-Canadian inventor of the telephone and the aileron on aircraft. Also, co-founder of the National Geographic Society.
Eric Temple Bell	(1883–1960)
Hilaire Belloc	(1870–1953), English writer.
Stephen Vincent Benet	(1898–1943), American poet.
Edmund Clerihew Bentley	(1875–1956), British writer.
Milton Berle	(1908–), American comedian and actor.
Daniel Bernoulli	(1700–1782), Dutch-born Swiss scientist who discovered basic principles of fluid behavior.
Joseph Bertrand	(1822–1900), French mathematician.
Ugo Betti	(1892–1953), Italian playwright.
Bhagavadgita	(100? C.E.), Epic Hindu poem forming part of the Hindu scripture. Title translates to "Song of God."
Patrick Blackett	(1897–1974), British physicist and Nobel laureate.
William Blake	(1757–1827), English poet and artist.
Niels Bohr	(1885–1962), Danish physicist who won the Nobel Prize for bringing quantum mechanics to the atomic model.
Janos Bolyai	(1802–1860), Hungarian mathematician who was one of the founders of non–Euclidean geometry.
Wolfgang Bolyai	(1775–1856), Father of Janos Bolyai.
Napoleon Bonaparte	(1769–1821), French emperor and great military commander.
Boorstein	Librarian of the Library of Congress.
James Boswell	(1740–1795), Scottish author and diarist.
Davis Bowles	Professor of civil and environmental engineering
Robert Boyle	(1627–1691), British scientist who separated chemistry from alchemy and was the first to isolate and study a gas.
Lawrence Bragg	(1890–1971), Australian-born British physicist who shared the Nobel Prize with his father for their use of X–rays to understand crystal structure.
Rafael Bras	Professor of civil engineering.
Bertolt Brecht	(1898–1956), German dramatist.
John Brennan	Engineer with the U.S. Army Corps of Engineers.
Jacob Bronowski	(1908–1974), Author and mathematician.
Emily Bronte	(1818–1848), English author.
Robert Browning	(1812–1889), English poet.
William Cullen Bryant	(1794–1878), American poet and politician.
Pearl S. Buck	(1892–1973), American novelist who won the Nobel Prize in literature.
Lois McMaster Bujold	Contemporary science-fiction author.
George Bush	(1924–), 41st president of the United States.
Ritchie Calder	(1898–), American engineer and sculptor.
David Caldwell	American cetacean biologist who co-discovered that dolphins use signature whistles.

Melba Caldwell	American cetacean biologist who co-discovered that dolphins use signature whistles.
Callimachus	(200s B.C.E.), Egyptian poet and librarian at the Library of Alexandria.
Thomas Carlyle	(1795–1881), Scottish essayist and historian.
Lord Carrington	(1919–), British statesman.
Lewis Carroll	(1832–1898), English author and mathematician.
Rachel Carson	(1907–1964), American marine biologist and author.
George Washington Carver	(1864–1943), American educator and innovator in agricultural science.
Willa Cather	(1873–1947), American writer.
William R. Catton, Jr	Professor of civil engineering and author of texts on hydrology.
Frank Chapman	Ornithologist and associate editor of National Geographic Magazine.
Charles, Prince of Wales	(1948–), Prince of England. His full name is Charles Philip Arthur George.
Agatha Christie	(1891–1976), British mystery writer.
Randolph Churchill	(1849–1895), British statesman.
Winston Churchill	(1874–1965), British Prime Minister during World War II.
Marcus Tillius Cicero	(106–43 B.C.E.), Roman writer and politician.
Jean Cocteau	(1889–1963), French novelist.
Samuel Taylor Coleridge	(1772–1834), English poet.
John Amos Comenius	(1592–1670), Czech educational reformer and religious leader.
Auguste Comte	(1798–1857), French positivist philosopher, who was a founder of sociology.
Joseph Conrad	(1857–1924), Polish-born English novelist.
Calvin Coolidge	(1872–1933), 30th president of the United States.
Gene L. Coon	Contemporary science fiction author.
Nicholaus Copernicus	(1473–1543), Polish astronomer who developed the heliocentric model of the solar system.
Edgar M. Cortright	Director of NASA's Langley Research Center.
Steven R. Covey	Author and motivational speaker.
Oliver Crawford	Contemporary science-fiction author.
Francis Crick	(1916–), British biophysicist, who won the Nobel Prize for helping to determine the structure of DNA.
Marie Curie	(1867–1934), Polish-born French scientist who won the Nobel Prize for the discovery of radioactive elements.
Leonardo da Vinci	(1452–1519), Italian artist and engineer.
Richard Henry Dana	(1815–1882), American writer and lawyer.
Dante	(1265–1321), Italian poet.
Charles Robert Darwin	(1809–1882), British scientist famous for his theory of evolution.
Honoré de Balzac	(1799–1850), Celebrated French author.
Pierre Augustin Caron de Beaumarchais	(1732–1799), French playwright and spy who sold weapons to Americans during the Revolution.

Lee De Forest	(1873–1961), American inventor of the vacuum tube who was the first to demonstrate sound in movies.
Edmond de Goncourt	(1822–1896), French writer who sought to portray history through an analysis of intimate, unpublished documents.
Jules de Goncourt	(1830–1870), Brother of Edmond who collaborated with Edmond's portrayal of history through unpublished documents.
Pierre-Simon de Laplace	(1749–1827), French astronomer and mathematician who helped to develop basic calculus.
Cecil B. De Mille	(1881–1959), American movie director and producer.
Gerard De Nerval	(1808–1855), French writer. Pseudonym of Gerard Larbrunie.
René Descartes	(1596–1650), French philosopher, scientist, and mathematician who developed the Cartesian coordinates.
Deuteronomy	Fifth book of the Hebrew Bible and the Christian Old Testament.
Irvine Devore	Behavioral and field biologist.
Charles Dickens	(1812–1870), English novelist.
Emily Dickinson	(1830–1886), America poet.
Paul Dirac	(1902–1984), British theoretical physicist who won the Nobel Prize for including relativistic considerations into the wave dynamics of electrons.
Benjamin Disraeli	(1804–1881), British writer and prime minister.
Charles Lutwidge Dodgson	Real name of Lewis Carroll.
Aelius Donatus	(circa 330 B.C.E.), Roman grammarian.
Donald Douglas	(1892–1981), American aeronautical engineer and aircraft manufacturer who designed the world's most popular passenger plane, the DC-3.
Arthur Conan Doyle	(1859–1930), British physician and novelist who created Sherlock Holmes.
René Dubos	(1901–1982), French-born American bacteriologist famous for his work with antibiotics.
Charles H. Duell	(1800s), Commissioner, U.S. Office of Patents
Dupont	Largest manufacturer of chemical products in the United States.
Will Durant	(1885–1981), American historian.
Freeman Dyson	(1923–), British-born American theoretical physicist and astrophysicist who helped to coordinate quantum mechanics with light.
Fred Eaton	Mayor of Los Angeles
Ecclesiastes	Book of the Hebrew Bible and the Christian Old Testament offering reflections on the purpose and nature of life.
Sir Arthur Stanley Eddington	(1882–1944), British astronomer and physicist.
Thomas Edison	(1847–1931), American inventor who developed the electric light bulb, electric generating system, sound-recording device, and motion picture projector.
A. D. Edmonston	California's state engineer.
Max Ehrlich	Contemporary science-fiction author.
Manfred Eigen	(1927–), German physical chemist who won the Nobel Prize for his studies of very rapid chemical reactions.

Albert Einstein	(1879–1955), German-born American physicist and Nobel laureate who devised the theories of special and general relativity.
Hans A. Einstein	American hydraulic engineer.
George Eliot	(1819–1880), English novelist. Pseudonym of Marian Evans.
Jan Eliot	Contemporary cartoonist.
Thomas Stearns Eliot	(1888–1965), American-born British poet who won the Nobel Prize for literature.
Elizabeth I, Queen	(1533–1603), Queen of England.
Duke Ellington	(1899–1974), American jazz composer and bandleader.
Havelock Ellis	(1859–1939), British sex psychologist and author.
Ralph Waldo Emerson	(1803–1882), American essayist, poet, and leader of the transcendentalism movement.
Quintus Ennius	(239–169? B.C.E.), Roman poet.
Epicurus	(341–270 B.C.E.), Greek philosopher, founder of Epicureanism, the basic premise of which is that intellectual pleasures are the supreme good.
Euclid	(300s B.C.E.), Greek mathematician, student of Plato, whose geometry book, the Elements, is still used today.
Leonhard Euler	(1707–1783), Swiss mathematician who helped found the field of advanced mathematics.
Jean Henri Fabre	(1823–1915), French entomologist noted for his studies of arthropod behavior.
Michael Faraday	(1791–1867), British physicist and chemist noted for his discoveries of electromagnetic induction and the laws of electrolysis.
Farmers' Almanac	(1848–), Agricultural and humor magazine.
Enrico Fermi	(1901–1954), Italian-American physicist who won the Nobel Prize for artificial radioactivity.
Richard Feynman	(1918–1988), American physicist and author who won the Nobel Prize for his theory of the conversion of light into matter.
Craig Fischenich	American research civil engineer.
Ronald Fisher	(1890–1962), British mathematician who showed that by partitioning the variations of a body of data, one can accurately assess how they influence one another and the outcome of an experiment.
D.C. Fontana	Contemporary science-fiction author.
Bernard Learning Bovier Fontenelle	(1657–1757), French dramatist and scientific writer.
Henry Ford	(1863–1947), American mechanical engineer who standardized the assembly line and founded the Ford Motor Company.
Joseph Ford	Professor and pioneer of chaos theory.
H.W. Fowler	(1859–1933), Author.
Francis Galton	(1822–1911), British scientist who pioneered the use of fingerprints for identification.
Victor Frankl	(1905–), Austrian psychotherapist and Holocaust survivor.
Ben Franklin	(1706–1790), American author, diplomat, and scientist.
Richard French	Professor of Civil Engineering.

Sigmund Freud	(1856–1939), Austrian physician, who founded psychoanalysis.
Archie Fripp, Jr.	(1939–), American microgravity scientist.
Archie Fripp, Sr.	(1906–1970), American army officer during WWII and Korea. He escaped from German captors through the use of a hand grenade at extremely close range and was wounded in the effort.
Michael Fripp	(1970–), American structural dynamicist and acoustician.
Max Frisch	(1911–1991), Swiss playwright.
Sextus Julius Frontinus	(40–130 C.E.), Water commissioner of Rome and highly regarded engineer.
Robert Frost	(1874–1963), American poet.
Buckminster Fuller	(1895–1983), American engineer noted for his use of technology to deal with problems facing humanity.
Solomon Ibn Gabirol	(1021?–1058?), Spanish Jewish philosopher and poet.
Galileo Galilei	(1564–1642), Italian physicist, and astronomer who perfected the telescope.
Evariste Galois	(1811–1832), French mathematician who helped to develop group theory.
Mohandas Gandhi	(1869–1948), Indian leader who led the country's independence movement, generally known as Mahatma Gandhi.
Cecilia Payne Gaposchkin	(1900–79), English-born American astrophysicist who was the first to measure chemical elements in stars by spectral analysis.
Martin Gardner	American scientific writer.
James Garfield	(1831–1881), 20th president of the United States.
Bill Gates	(1955–), American business executive and co-founder of the Microsoft Corporation.
Karl Friedrich Gauss	(1777–1855), German mathematician and physicist. He did not publish much of his work in order to avoid publicity.
Genesis	First book of the Hebrew Bible and Old Testament which describes the creation of the world.
David Gerrold	Contemporary science-fiction author.
W. S. Gilbert	(1836–1911), English playwright.
Charlotte Perkins Stetson Gilman	(1860–1935), American feminist and writer.
Jean Giraudoux	(1882–1944), French diplomat and author.
Ellen Glasgow	(1873–1945), American novelist and southerner who wrote realistic accounts of the antebellum south.
Maria Goeppert-Mayer	(1906–1972), German-American physicist who won the Nobel Prize for her study of the atomic nucleus.
Johann Goethe	(1749–1832), German writer and scientist who wrote Faust and studied comparative morphology.
Daniel S. Goldin	Administrator of NASA.
George Gordon	(1788–1824), English poet known as Lord Byron.
Al Gore	(1948–), American senator and the 45th vice president of the United States.
Steven J. Gould	(1941–), American professor and author. Proposed the "punctuated equilibrium" theory of evolution which holds that rather than evolving gradually, species evolve rapidly over short periods of time, then remain stable for long periods afterward.

Kenneth Grahame	(1859–1932), English author.
Robert Graves	(1895–1985), British poet and novelist.
Matt Groening	(1954–), American cartoonist who created *The Simpsons*.
Haida	Group of tribes native to the Queen Charlotte Islands in British Columbia, Canada.
John Haldane	(1892–1964), British geneticist who mathematically modeled rates of genetic variation.
W.D. Hamilton	British biologist who developed the concept of kin selection which holds that animals can best enhance their own reproductive success by aiding their close relatives so long as the gain conferred on the recipient is greater than the cost to the donor.
Dag Hammarskjold	(1905–1961), Swedish statesman, UN secretary general, and Nobel Peace Prize winner.
Godfrey Hardy	British mathematician who helped develop the model for genetic variations in a population.
Harry Harlow	American experimental psychologist best known for his experiments on the effects of social isolation in monkeys.
Al Harrison	Fluid dynamist and expert in sedimentation problems of the Missouri River.
William Harvey	(1578–1657), English physician, who discovered the circulation of the blood and the role of the heart in propelling it.
Stephen Hawking	(1942–), British theoretical physicist, best known for his attempts to unite general relativity theory with quantum mechanics and for his work on black holes.
Nathaniel Hawthorne	(1804–1864), American author.
John Hay	(1838–1905), American writer and secretary of state.
Oliver Heaviside	(1850–1925), British physicist who predicted the existence of the ionosphere.
Robert A. Heinlein	(1907–1988), American science-fiction author.
Werner Heisenberg	(1901–1976), German physicist and Nobel laureate who investigated quantum mechanics.
Joseph Heller	(1923–), American novelist.
Ernest Hemingway	(1899–1961), American novelist.
Heraclitus	(540?–475? B.C.E.), Greek philosopher who believed that fire is the primordial source of matter.
Herodotus	(484?–425 B.C.E.), Greek historian known as the father of history.
Heinrich Hertz	(1857–94), German physicist who proved that electricity can travel in electromagnetic waves.
David Hilbert	(1862–1943), German mathematician who generalized Euclid's models of geometry.
L. Erskine Hill	Lecturer in physiology at London Hospital.
Robert Hinde	American behavioral biologist.
Adolph Hitler	(1889–1945), Austrian-born Nazi dictator of Germany.
Eric Hoffer	(1902– 1983), Philosopher and longshoreman.
Oliver Wendell Holmes	(1809–1894), American writer and physician.
Horace	(65–8 B.C.E.), Roman lyric poet.

Fred Hoyle	(1915–), English astronomer and author who was one of the first to apply relativity to cosmology.
William Henry Hudson	(1841–1922), English writer and naturalist specializing in ornothology.
Victor Hugo	(1802–1885), French novelist.
Zora Neale Hurston	(1901?–1960), American writer and folklorist of black culture.
Charlie Husson	Engineer with NASA .
Aldous Huxley	(1894–1963), English novelist.
Thomas Huxley	(1825–1895), British biologist, best known for his active support of Charles Darwin's theory of evolution.
Christiaan Huygens	(1629–95), Dutch astronomer, mathematician who developed the wave theory of light and the pendulum clock.
Joris Karl Huysmans	(1848–1907), French novelist. Real name is Charles Huysmans.
Lee Iacocca	(1924–), American corporate executive.
John James Ingalls	(1833–1900), American Senator
Andrew Jackson	(1767–1845), Seventh president of the United States.
Carl Jacobi	(1804–1851), German mathematician who was one of the founders of the theory of elliptic functions.
William James	(1842–1910), American philosopher and psychologist.
James Jeans	(1877–1946), British mathematician, physicist, and astronomer who applied mathematics to problems in cosmology.
Thomas Jefferson	(1743–1826), Third president of the United States and author of the Declaration of Independence.
Jeremiah	Book of the Hebrew Bible and Christian Old Testament about the prophecies of Jeremiah, (650–580 B.C.E.), and the Babylonian exile.
Job	Book of the Hebrew Bible and Christian Old Testament describing the suffering of the righteous at the hands of God.
Pope John Paul II	(1920–), Polish pope. Born Karol Wojtyla.
Lyndon B. Johnson	(1908–1973), 36th president of the United States.
Samuel Johnson	(1709–1784), English writer and author of a dictionary.
Benjamin Jowett	(1817–1893), British educator and Greek scholar.
Juvenal	(65–128 C.E.), Roman satirical poet. Full name is Decimus Junius Juvenalis.
Leo Kadanoff	(1900s) American physicist studying the phase transition in metal.
Stephen Kandel	Contemporary science-fiction author.
Emmanual Kant	(1724–1804), German philosopher.
Samuel Karlin	(1923–), Mathematics professor who studies the methodology of operations research and management science.
John Keats	(1795–1821), English poet.
Helen Keller	(1880–1968), American author and lecturer who was deaf and blind.
William Thomson Kelvin	(1824–1907), British mathematician, physicist, and educator whose discoveries ranged from temperature scales to telegraph cables.
John F. Kennedy	(1917–1963), 35th president of the United States.

Johannes Kepler	(1571–1630), German astronomer who formulated and verified the laws of planetary motion.
Charles F. Kettering	(1876–1958), American engineer who invented the automatic starter for automobiles.
John Keynes	(1883–1946), British economist.
Ibn Khaldun	(1332–1406), Islamic historian.
Joyce Kilmer	(1886–1918), Poet who died in WW I.
Rudyard Kipling	(1865–1936), English writer who won the Nobel Prize in literature.
Henry Kissinger	(1923–), American secretary of state and Nobel laureate for peace in Vietnam.
Arthur Koestler	(1905–1983), English novelist.
Koran	Sacred scripture of Islam containing Mohammed's revelations.
Joseph Wood Krutch	(1893–1970), American educator and naturalist.
Thomas Kuhn	(1922–1996), American historian and philosopher of the development of science.
Lucy Larcom	(1826–1893), American poet.
Fiorello La Guardia	(1882–1947), Mayor of New York City.
Joseph Louis Lagrange	(1736–1813), French mathematician who developed calculus of variations and the metric system.
Lao–Tzu	(570?–490? B.C.E.), Chinese philosopher and founder of Taoism.
D.H. Lawrence	(1885–1930), English novelist and poet who was critical of industrialization.
Ursula K. Le Guin	(1929–), American science-fiction writer.
Stephen Leacock	(1869–1944), Canadian writer and economist.
William Daniel Leahy	(1875–1959), American admiral and diplomat.
Stephen Leatherwood	(–1998), American cetacean biologist.
Henri Lebesgue	(1875–1941), French mathematician who generalized the definition of the definite integral.
Arnold Lehman	Member of the U.S. Food and Drug Administration.
Gottfried Wilhelm Leibnitz	(1646–1716), German philosopher, mathematician, and statesman.
John Lennon	(1940–1980), British singer, songwriter and member of the Beatles.
Aldo Leopold	(1886–1948), American naturalist who pioneered the application of environmental principles to wildlife management.
Luna Leopold	Geologist, civil engineer, and author of texts on geomorphology.
Albert Libchaber	French physicist studying the superfluid behavior of helium
George Christoph Lichtenberg	(1742–1799), German physicist and satirical writer who discovered the basic principle of modern xerographic copying.
Jon Lien	Canadian cetacean biologist.
John C. Lilly	Disreputed American cetacean biologist and author.
Charles Lindbergh	(1902–1974), American aviator who made the first solo, nonstop, transatlantic flight in 1927.
Carolus Linnaeus	(1707–1778), Swedish naturalist who developed binomial nomenclature to classify and organize plants and animals.

Gabriel Lippman	(1845–1921), French physicist who won the Nobel Prize for his invention of the process for making color photographs that did not fade quickly after development.
J.E. Littlewood	(1885–1977), English mathematician.
John Locke	(1632–1704), English philosopher.
Jack London	(1876–1916), American writer.
Anita Loos	(1888–1981), American novelist and screenwriter.
Konrad Lorenz	(1903–89), Austrian zoologist and Nobel Laureate for his work on ethology.
Thomas E. Lovejoy	Tropical ecologist and the Assistant Secretary for External Affairs at the Smithsonian Institution
Lucretius	(98?–55? B.C.E.), Roman poet.
Luke	Third book of the New Testament.
Martin Luther	(1483–1546), German theologian who initiated the Protestant Reformation.
Edward Lytton	(1803–1873), English novelist and politician.
Maccabees	Book in Hebrew Bible and Christian Old Testament dealing with early Jewish history.
Ernst Mach	(1838–1916), Austrian physicist who performed important work in the aerodynamics of ballistics.
Archibald MacLeish	(1892–1982), American poet and librarian of congress between 1929–1944.
David Maidment	Professor of civil engineering.
Thomas Mann	(1875–1955), German novelist who won the Nobel Prize in literature.
Edwin Markham	(1852–1940), American poet.
Karl Marx	(1818–1883), German cofounder of communism.
Matthew	First book of the New Testament of the Christian Bible.
James Maxwell	(1831–1879), British physicist whose theories unified electricity and magnetism.
Larry Mays	Professor of civil engineering.
Jim McGlincy	American sailor.
Marshall McLuhan	(1911–1980), Canadian writer who believed that the electronic media would have a strong effect on culture.
Margaret Mead	(1901–78), American anthropologist.
Herman Melville	(1819–1891), American novelist.
Menaechmus	(300s B.C.E.), Greek mathematician who developed the theory of conic sections.
Mickey Mouse	(1928–), Cartoon character created by Walt Disney.
John Stuart Mill	(1806–1873), English philosopher, political economist, and exponent of Utilitarianism.
Robert A. Millikan	(1868–1953), American physicist who won the Nobel Prize for measuring the charge on an electron.
John Milton	(1608–1674), English poet.
Margaret Mitchell	(1900–1949), American author.
Maria Mitchell	(1818–89), American astronomer and professor.
Russell Mittermeir	Primatologist and president of Conservation International.

Claude Monet	(1840–1926), French impressionist painter.
Jacques Monod	(1910–1979), French biochemist who won the Nobel Prize in Medicine.
Charles Louis de Secondat Montesquieu	(1689–1755), French writer and jurist.
Ronald D. Moore	Contemporary science-fiction author.
Lewis Mumford	(1895–1990), American social philosopher and urban planner.
Hector Hugh Munro	(1870–1916), British writer also known as "Saki."
NASA	(1958–), National Aeronautics and Space Administration.
Navajo	Native American tribe living in Utah and New Mexico.
Charles Neaves	(1800–1876), Scottish poet.
NEPA	National Environmental Policy Act. NEPA is the basic national charter for the protection of the environment. The charter ensures that the environmental impact of proposed federal projects is considered.
Homer E. Newell	Associate Administrator of NASA.
Isaac Newton	(1642–1727), English mathematician and physicist who developed calculus, formulated the laws of motion and gravitation, and solved the problems of optics.
Friedrich Nietzsche	(1844–1900), German philosopher.
Florence Nightingale	(1820–1910), British nurse, hospital reformer, and humanitarian.
Leonard Nimoy	(1931–), Contemporary science-fiction actor and author best known as Spock on Star Trek.
Ken Norris	(1924 –1998), American cetacean biologist who discovered that dolphins use sonar.
H. Pierre Noyes	Physics professor.
Flannery O'Connor	(1925–1964), American writer.
Sorai Ogyu	(1666–1729), Japanese scholar of Chinese culture.
Bori L. Olla	Cetacean biologist.
Ken Olson	President and founder of Digital Equipment Corporation.
Jose Ortega y Gasset	(1883–1955), Spanish essayist and philosopher.
George Orwell	(1903–1950), British writer. Real name is Eric Blair.
Fairfield Osborn	(1857–1935), American paleontologist.
Sir William Osler	(1849–1919), Canadian physician and medical researcher.
Paracelsus	(1493–1541), German physician and alchemist. Pseudonym of Theophrastus Bombastus von Hohenheim.
Vilfredo Pareto	(1848–1923), Italian sociologist and economist.
Blaise Pascal	(1623–1662), French mathematician and physicist.
Louis Pasteur	(1822–1895), French biologist who founded microbiology, invented the process of pasteurization, and developed vaccines for several diseases, including rabies.
General Patton	(1885–1945), American World War II army general.
Epistles of Paul	Books of the New Testament establishing the doctrine and organization of the Christian Church.
Linus Pauling	(1901–1994), American chemist and physicist who won the Nobel Prize for genetic chemistry and won the Nobel Peace Prize.
Karl Pearson	(1857–1936), British mathematician who applied statistics to questions of heredity.

Joachim Peiper	German Panzer tank division commander in WWII.
Mike Peters	Contemporary cartoonist.
Benjamin Pierce	(1809–1880)
Arthur Pillsbury	American hydraulic engineer.
Pius XII	(1876–1958), Pope during World War II.
Max Planck	(1858–1947), German physicist who won the Nobel Prize for originating quantum theory.
Plato	(428?–347 B.C.E.), Greek philosopher.
Plotinus	(205–270 C.E.), Roman philosopher.
Plutarch	(46?–120 C.E.), Greek biographer and essayist.
Pierre Pochet	Professor of Physiology at Toulouse, France.
Henri Poincaré	(1854–1912), French physicist and mathematician.
Simeon Poisson	(1781–1840), French mathematician and physicist who did extensive work in elasticity, statistics, and calculus.
George Polya	(1887–1985), Hungarian-born American mathematician.
Karl Popper	(1902–1994), Austrian-British philosopher of science.
George Porter	(1920–), British chemist who won the Nobel Prize in chemistry for his use of photolysis to study chemical reactions.
Chaim Potok	Contemporary Jewish author.
Colin Powell	(1937–), American military leader and army general.
William Preece	(1800s), Chief engineer of the British Post Office.
Stuart Prinam	Conservation biologist.
Proverbs	Book of the Hebrew Bible and the Christian Old Testament containing expressions of wisdom and experience.
Karen Pryor	American cetacean biologist and dolphin trainer.
Psalms	Book of the Hebrew Bible and the Christian Old Testament containing 150 songs and poems.
E. Purcell	(1912–1997), American physicist and educator who won the Nobel Prize for developing nuclear magnetic resonance.
Adolphe Quetelet	Belgian astronomer who in 1835 was the first to note the periodic nature of the Perseids meteor showers.
Ayn Rand	(1905–1982), American novelist and philosopher.
Ronald Reagan	(1911–), 40th president of the United States.
Edward Redish	(1942–), American physicist researching proper techniques for teaching physics.
Randall Reeves	American cetacean biologist.
Ernest Renan	(1823–92), French philologist and historian of religion.
Grantland Rice	(1880–1954), American sportswriter.
Lewis F. Richardson	English fluid dynamicist.
Edwin Arlington Robinson	(1869–1935), American poet.
Laurance S. Rockefeller	(1910–), American philanthropist and conservationist who established the Virgin Islands National Park.
Gene Roddenberry	(1921–1991), American television scriptwriter and creator of Star Trek.
Wilhelm Roentgen	(1845–1923), German physicist who won the Nobel Prize for his discovery of X-rays.
Will Rogers	(1879–1935), American humorist and actor.

Robert Romer	Contemporary physicist.
Andy Rooney	American humorist.
Eleanor Roosevelt	(1884–1962), social activist, United Nations representative and wife Franklin D. Roosevelt.
Franklin Delano Roosevelt	(1882–1945), 32nd president of the United States.
Theodore Roosevelt	(1858–1919), 26th president of the United States.
Jean-Jacques Rousseau	(1712–1778), French writer and proponent of the "noble savage," the theory that the primitive state is morally superior to the civilized state.
F. Sherwood Rowland	(1927–), American chemist and Nobel Prize winner who demonstrated the effect of CFCs on the ozone layer.
William D. Ruckelshaus	(1932–), American lawyer and first head of the Environmental Protection Agency.
Bertrand Russell	(1872–1970), British author and mathematician who won the Nobel Prize for literature.
Ernest Rutherford	(1871–1937), British physicist who won the Nobel Prize for his theory of the structure of the atom.
William Safire	(1927–), American political columnist.
Carl Sagan	(1934–1996), American astronomer and author.
Andrei Sakharov	(1921–1989), Soviet physicist and political dissident who worked on the development of the Soviet hydrogen bomb.
Robert Sapolsky	American biologist who studies the effects of stress hormones upon mammalian physiology and behavior.
Sappho	(610?–580? B.C.E.), Greek poet.
David Sarnoff	(1891–1971), Russian-born American broadcasting executive.
John Ralston Saul	Contemporary author.
Charles Melville Scammon	Early whaler-naturalist.
George B. Schaller	American behavioral biologist known for his studies of gorillas in their natural habitat.
Glenn Seaborg	(1912–), American chemist who won the Nobel Prize for his characterization of radioactive isotopes.
Jerry Seinfeld	American comedian and actor.
Hans Selye	Canadian physician who characterized the response to stress.
Seneca	(4? B.C.E.–65 C.E.), Roman writer.
Dr. Seuss	(1904–1991), American author of children's books. Real name is Theodore Seuss Geisel.
William Shakespeare	(1564–1616), English playwright and poet.
George Bernard Shaw	(1856–1950), Irish writer.
Percy Bysshe Shelley	(1792–1822), English poet.
Eugene M. Shoemaker	Chief of the Astrogeology Branch of the U.S. Geological Survey.
Armin Shoklitsch	Professor of hydraulic engineering.
Igor Sikorsky	(1889–1972), Russian-born American aeronautical engineer who developed the first American helicopter.
Simplicius	(500s B.C.E.), Greek natural philosopher.
The Simpsons	(1990–), Animated comedy television show created by Matt Groening.
Henry John Stephen Smith	(1826–1883)

Sydney Smith	(1771–1845), English writer and Anglican clergyman.
Socrates	(470?–399? B.C.E.), Greek philosopher.
King Solomon	King of ancient Israel (reigned 961–922 B.C.E.) and considered one of the wisest sages.
Susan Sontag	(1913–), American essayist.
Sophocles	(496?–406? B.C.E.), Greek playwright.
Herbert Spencer	(1820–1903), British sociologist and evolutionist who coined the phrase "survival of the fittest".
Stephen Spender	(1909–1995), English poet.
St. Paul's Church	(1692–), Church in Baltimore, Maryland.
Star Trek	(1966–1969), Science-fiction television drama created by Gene Roddenberry.
Star Trek, The Next Generation	(1987–1994), Science-fiction television drama created by Gene Roddenberry.
Caecilius Statius	(220–168 B.C.E.), Roman comic playwright.
Gertrude Stein	(1874–1946), American writer and patron of the arts.
David B. Steinman	Bridge engineer credited with greatly increasing the understanding of aerodynamic bridge design.
Charles Proteus Steinmetz	(1865–1923), German-born American electrical engineer who developed the theory describing hysteresis and alternating current circuits.
Charles P. Stevens	Geology professor.
Wallace Stevens	(1879–1955), American poet.
Adlai Stevenson	(1900–1965), U.S. statesman and diplomat.
Robert Lewis Stevenson	(1850–1894), Scottish novelist.
Potter Stewart	(1915–85), associate justice of the U.S. Supreme Court.
Sting	(1951–), British rock musician whose real name is Gordon Sumner.
William Stolzenburg	Associate editor of Nature Conservancy.
Tom Stoppard	(1937–), English playwright.
Michael W. Straus	Commissioner of the Bureau of Reclamation.
Lewis Strauss	Chairman of the Atomic Energy Commission.
Geoffrey Streatfield	(1897–1978), British lawyer.
Simeon Strunsky	(1879–1948), Book reviewer.
Gordon Matthew Sumner	(1951–), British rock musician known as Sting.
Swamp Land Act	(1849), Nation's first federal wetland policy which encouraged the conversion of wetlands into farms.
Albert Szent-Gyorgyi	(1893–), Hungarian-American scientist.
William Taft	(1857–1930), 27th president and 10th chief justice of the United States.
Rabindranath Tagore	(1861–1941), Indian poet and Nobel laureate for literature.
Edward Teller	(1908–), Hungarian-American physicist known for his work on the atomic and hydrogen bombs.
Nikola Tesla	(1856–1943), Serbian-American electrical engineer and inventor of the transformer and the AC motor.
Thales	(625?–546? BC), Greek philosopher who introduced geometry to Greece and who believed that the original principle of all things is water, from which everything proceeds and into which everything is again resolved.

Margaret Thatcher	(1925–), British Prime Minister.
Bob Thaves	Contemporary cartoonist.
Theognis	(570?–490? B.C.E.), Greek poet.
Spinal Tap	(1984), Satirical rock movie.
David Hurst Thomas	Contemporary archaeologist and textbook author.
Lewis Thomas	(1913–1993), American physician and author.
Henry David Thoreau	(1817–1862), American writer who celebrated nature and individualism.
James Thurber	(1894–1961), American cartoonist and author.
Gherman Titov	(1935–), Soviet cosmonaut and the second person to orbit the earth.
Isaac Todhunter	(1820–1884), British professor, author, and historian of mathematics.
Leo Tolstoy	(1828–1910), Russian novelist.
George Macaulay Trevelyan	(1876–1962), British historian.
Leon Trotsky	(1879–1940), Russian Marxist and leader in the Russian Revolution.
Harry S. Truman	(1884–1972), 33rd president of the United States.
Konstantin Tsiolkovsky	Russian schoolteacher who first proposed liquid rocket propellant.
R. Turner	American cetacean biologist.
Alan Turning	(1912–1954), One of the early creators of computers, credited with using a rudimentary computer to break the German military codes in World War II.
Mark Twain	(1835–1910), American writer and humorist.
John Tyndall	(1820–1893), British physicist, noted for his study of colloids.
Stewart Udall	U.S. Congressman and Secretary of the Interior.
John Updike	(1932–), American writer.
Vincent van Gogh	(1853–1890), Dutch postimpressionist painter.
Anton van Leeuwenhoek	(1632–1723), Dutch inventor of one of the first microscopes who made pioneering discoveries concerning protozoa, red blood cells, and the life cycles of insects.
Thorstein Veblen	(1857–1929), American economist and social scientist.
Charles M. Vest	American mechanical engineer and president of the Massachusetts Institute of Technology.
Virgil	(70–19 B.C.E.), Roman poet.
Virginia Wetland Act	(1972), Law encouraging the preservation of wetlands.
Voltaire	(1694–1778), French writer and philosopher. Real name is Francois Marie Arouet.
Wernher von Braun	(1912–1977), German-American engineer who developed the liquid-fuel rocket.
Alexander von Humboldt	(1769–1859), German naturalist and explorer of the American west.
Theodore von Kármán	(1881–1963), Hungarian-born American aeronautical engineer who helped to develop fluid mechanics, turbulence, and supersonic flight.

Johann von Neumann	(1903–1957), Hungarian-American mathematician who developed the branch of mathematics known as game theory.
Frances A. Walker	Superintendent of the 1870 Census.
Izaak Walton	(1593–1683), English author and sport fisherman.
Sylvia Warner	(1893–1978), English novelist.
George Washington	(1732–1799), First president of the United States.
James D. Watson	(1928–), American biochemist and Nobel laureate who helped to determine the structure of DNA.
Lyall Watson	Contemporary South African biologist and anthropologist.
Thomas Watson	(1874–1956), American engineer and founder of IBM.
Simone Weil	(1909–1943), French social philosopher.
Julius Weisbach	(1800s), German mechanical engineer and author.
William H. Welch	(1850–1934), American physician
Arthur Mellen Wellington	(1847–1895), American engineer.
H.G. Wells	(1866–1946), British writer.
James Whistler	(1834–1903), American painter.
E. B. White	(1899–1985), American writer.
Alfred North Whitehead	(1861–1947), British mathematician who philosophized upon the theory of logic.
Christene Todd Whitman	Governor of New Jersey.
Walt Whitman	(1819–1892), American poet.
Norbert Wiener	American mathematician and founder of cybernetics, the study of control and communication in machines, animals, and organizations.
Peter Wilcock	Professor of Geography and Environmental Engineering.
Tennessee Williams	(1911–1983), American playwright.
Edward O. Wilson	(1929–), Biologist and author on biodiversity and comparative zoology.
Woodrow Wilson	(1856–1924), 28th president of the United States.
Howard E. Winn	American cetacean biologist.
Ludwig Wittgenstein	(1889–1951), Austrian-British philosopher.
Tom Wolfe	Contemporary American novelist.
William Wordsworth	(1770–1850), British poet.
Frank Lloyd Wright	(1867–1959), American architect.
Orville Wright	(1871–1948), American aeronautical engineer who developed the first airplane with his brother Wilbur.
Wilbur Wright	(1867–1912), American aeronautical engineer who developed the first airplane with his brother Orville.
Malcolm X	(1925–1965), Black American civil rights leader. Original name is Malcolm Little and his Muslim name is El–Hajj Malik El–Shabazz.
Xenophanes	(580– B.C.E.), Greek poet, philosopher, and religious reformer.
Yevgeny Yevtushenko	Contemporary Russia poet.
Emperor Yu of China	(2000s B.C.E.), Legendary wise emperor who was deeply committed to flood control projects. One of the first emperors of China.
Zeuxis	(300s B.C.E.), Greek painter.

Index

A

Abbey, Edward, 141, 217
Abbott, Donald, 16, 17, 19, 22, 217
Abelard, Pierre, 179, 217
Ackoff, Russell, 185, 217
Adams, Abby, 144
Adams, John, 181, 217
Adams, Mike, 13
Adler, Alfred, 73, 217
Adler, Jerry, 124
Agassiz, Jean, 207, 217
Ahmes the Scribe, 54
Alcock, J., 24, 217
Alighieri, Dante, 146, 219
Allen, J., 106, 153, 154
Allen, Tim, 91, 217
Alsobrook, David, 183, 217
Altmann, Jeanne, 21, 217
Anglin, W., 48
Arbiter, Petronius, 190, 217
Archimedes, 72, 78, 105, 217
Aristotle, 55, 66, 122, 159, 217
Armstrong, Neil, 84, 217
Army Corps of Engineers, 106, 111, 138, 213, 217
Aroeste, Jean, 183
Ashurnasirpal II, 163, 217
Asimov, Isaac, 32, 94, 127, 217
Augarten, Stan, 95
Augustine, Saint, 10, 68, 177, 217
Aurelius, Marcus, 138, 217
Avery, Tex, 9
Ayers, Jeff, 28

B

Babbage, Charles, 97, 217
Babylonian Talmud, 134, 217
Bacon, Roger, 44, 62, 217
Bagehot, Walter, 49, 217
Bahu, Parakrama, 108, 217
Bailey, Liberty, 149, 217
Bain, M., 106, 153, 79

Balzac, Honore de, 207, 219
Banks-Smith, Nancy, 20, 217
Barnerd, James, 159, 217
Bartholemew, G., 161
Bateson, Patrick, 41, 217
Bauer, H., 76
Beaumarchais, Pierre de, 197, 219
Beebe, Spencer, 135, 217
Behr, Steven, 207, 217
Bell, Alexander, 78, 79, 188, 218
Bell, Eric, 48, 54, 66, 80, 218
Belloc, Hilaire, 218
Ben Tre, 138
Benet, Stephen, 84, 218
Benitez, Sandra, 80
Bennett, Harve, 83
Bentley, Edmund, 16, 218
Berger, 137
Berle, Milton, 181, 218
Bernoulli, Daniel, 68, 218
Bertrand, Joseph, 50, 218
Betti, Ugo, 147, 218
Bhagavadgita, 155, 201, 208, 218
Bingham, H., 21
Blackett, Patrick, 34, 218
Blake, Henry, 201
Blake, William, 57, 59, 63, 91, 218
Bohr, Niels, 4, 7, 11, 54, 177, 218
Bolyai, Janos, 71, 218
Bolyai, Wolfgang, 71, 218
Bonaparte, Napoleon, 112, 200, 218
Bordas-Demoulins, 65
Boorstein, Daniel, 179, 218
Boston Post, 118
Boswell, James, 161, 218
Bowles, Davis, 110, 218
Boyle, Robert, 40, 218
Bragg, Lawrence, 11, 218
Bras, Rafael, 105, 131, 218
Braun, Wernher von, 31, 85-88, 94, 192, 231
Brecht, Bertolt, 2, 218
Brennan, John, 108, 218
Brennan, Louis, 29
Brenner, Sydney, 97
Brent, Robert, 14

Brew, J., 40
Bronowski, Jacob, 6, 126, 218
Bronte, Emily, 151, 218
Browning, Robert, 145, 218
Bryant, William, 146, 150, 218
Buchholz, R., 22, 232
Buck, Pearl, 73, 218
Bugs Bunny, 9
Bujold, Lois, 76, 218
Buller, Arthur, 8
Burn, Barbara, 147, 148
Burns, John, 154
Burns, Judy, 194
Bush, George, 135, 218
Butler, Samuel, 81
Byers, W., 229
Byrd, William, 151

C

Caesar, Augustus, 100, 201, 217
Calder, Ritchie, 92, 218
Caldwell, David, 162, 218
Caldwell, Melba, 162, 219
Calivino, Italo, 96
Callimachus, 171, 219
Carlyle, Thomas, 178, 219
Carrington, Lord, 182, 219
Carroll, Lewis, 24, 67, 124, 192, 219
Carson, Rachel, 3, 37, 14, 134, 145, 165, 195, 219
Cartmill, Matt, 39
Carver, George, 148, 219
Cather, Willa, 150, 219
Catton, William, Jr., 17, 219
Chapman, Frank, 163, 219
Charles, Prince of Wales, 141, 219
Chien-Shiung Wu, 226
Chow, Ven, 132
Christie, Agatha, 29, 219
Churchill, Randolph, 70, 219
Churchill, Winston, 172, 179, 219
Cicero, Marcus, 99, 100, 178, 219
Clemmons, J., 22, 68
Cocteau, Jean, 197, 219
Coleridge, Samuel, 129, 130, 219
Collier, Abram, 158
Comenius, John, 179, 184, 219
Comte, Auguste, 14, 219
Conduit, Carl, 111
Conrad, Joseph, 128, 219
Cook, Michael, 74
Cook, Rich, 95
Coolidge, Calvin, 203, 219
Coon, Gene, 191, 219
Copernicus, Nicholaus, 70, 219
Cortright, Edgar, 88, 219
Covey, Steven, 188, 196, 219

Cowen, Richard, 20, 30, 149, 166
Crawford, Oliver, 25, 219
Creagh, Patrick, 96
Crick, Francis, 17, 72, 219
Cudmore, Larison, 35, 37
Curie, Marie, 36, 37, 219

D

da Vinci, Leonardo, 52, 56, 61, 62, 77, 82, 97, 104, 113, 144, 197, 201, 204, 219
Daly, Herman, 140
Dana, Richard, 159, 219
Dante Alighieri, 146, 219
Darwin, Charles, 18, 24, 33, 44, 166, 219
Davies, William Henry, 127
Davis, Jonathan, 29
Davis, Philip, 55
de Balzac, Honore, 207, 219
de Beaumarchais, Pierre, 197, 219
De Forest, Lee, 87, 219
de Goncourt, Edmond, 7, 220
de Goncourt, Jules, 7, 220
de Laplace, Pierre-Simon, 201, 220
De Mille, Cecil, 211, 220
de Nerval, Gerard, 148, 220
DeBin, Jerry, 125
Descartes, Rene, 32, 69, 200, 220
Desch, Eric, 16
Deuteronomy, 134, 220
Devore, Irvine, 22, 220
Dickens, Charles, 165, 173, 220
Dickinson, Emily, 145, 220
Dirac, Paul, 60, 62, 171, 220
Disraeli, Benjamin, ii, 24, 183, 220
Dobie, J. Frank, 4, 184
Dodgson, Charles, 220
Donatus, Aelius, 170, 220
Doran, James, 54
Douglas, Donald, 83, 220
Doyle, Arthur, 33, 60, 104, 172, 220
Drabek, R., 45
Dubos, Rene, 136, 220
Duell, Charles, 118, 220
Duffey, R.N., 223
Dumas, Alexandre, 57
DuPont, 153, 220
Durand, W., 105
Durant, Will, 138, 220
Dyson, Freeman, 12, 63, 76, 220

E

Eagleson, P., 105
Earle, Sylvia, 157, 158
Eaton, Fred, 108, 220
Ecclesiastes, 35, 107, 155, 202, 220

Eddington, Sir Arthur, 5, 47, 54, 220
Edison, Thomas, 112, 202, 220
Edmonston, A., 09, 220
Egrafor, M., 73
Ehrlich, Max, 166, 220
Eigen, Manfred, 52, 220
Einstein, Albert, 2, 4, 7, 8, 61, 55, 58, 63, 64, 79, 91, 92, 135, 144, 154, 170, 181, 22
Einstein, Hans, 221
Eisenhower, Dwight, 190
Eliot, George, 109, 165, 171, 221
Eliot, Jan, 171, 221
Eliot, Thomas, 156, 221
Elizabeth I, Queen, 157, 221
Ellerbee, Linda, 211
Ellington, Duke, 202, 221
Ellis, Havelock, 32, 67, 95, 102, 125, 221, 221
Emerson, Ralph, i, 14, 21, 39, 53, 89, 90, 127, 145, 146, 148, 149, 164, 183, 192, 198, 221, 242
Ennius, Quintus, 23, 87, 221
Epicurus, 208, 221
Erdos, Paul, 67
Euclid, 54, 71, 223
Euler, Leonhard, 72, 221

F

Fabre, Jean, 30, 220
Faraday, Michael, 14, 89, 90, 221
Fermi, Enrico, 14, 89, 90, 221
Feynman, Richard, 5, 7, 15, 53, 68, 144, 170, 171, 175, 221
Fischenich, Craig, 34, 189, 221
Fisher, Ronald, 26, 60, 221
Flood Control Act, 110
Fontana, D., 94, 221
Fontenelle, Bernard, 72, 221
Ford, Henry, 124, 221
Ford, Joseph, 4, 25, 221
Forrester, J.W., 125
Fowler, F.G., 173
Fowler, H., 172, 173, 221
Fownes, George, 10, 13, 77
Frankl, Victor, 203, 221
Franklin, Benjamin, 131, 221
Freeman, John, 61
French, Richard, 104, 221
French, Thomas, 80
Freud, Sigmund, 55, 195, 221
Fripp, Archie, Jr., 147, 221
Fripp, Archie, Sr., 201, 221
Fripp, Michael, 12, 100, 102, 222
Frisch, Max, 118, 222
Frontinus, Sextus, 118, 129, 222
Frost, Robert, 176, 221
Fuller, Buckminster, 54, 222

G

Gabirol, Solomon, 197, 222
Gale, Henry, 10, 92
Galilei, Galileo, 9, 57, 178, 222
Galois, Evariste, 171, 222
Galton, Francis, 28, 51, 222
Gandhi, Mohandas, 194, 200, 222
Gaposchkin, Cecilia, 39, 222
Gardner, Martin, 30, 64, 222
Garfield, James, 110, 222
Gaskin, D., 22, 162
Gates, Bill, 82, 94–96, 222
Gauss, Karl, 53, 112, 170, 222
Geisel, Theodore, 195, 222
Gekko, Gordon, 208
Genesis, 122, 127, 222
George, William, 192
Gerrold, David, 164, 222
Gilbert, Harold, 83
Gilbert, William, 74, 222
Gilman, Charlotte, 83, 222
Giraudoux, Jean, 138, 222
Glasgow, Ellen, 116, 222
Glassie, Henry, 100
Glegg, Gordon, 76
Gleick, James, 2, 144
Goeppert-Mayer, Maria, 64, 222
Goethe, Johann, 33, 40, 44, 119, 126, 222
Gogh, Vincent van, 145, 232
Goldin, Daniel, 83, 221
Goncourt, Edmond de, 7, 28, 220
Goncourt, Jules de, 8, 19, 220
Gordon, George, 9, 146, 157, 222
Gore, Al, 108, 123, 124, 142, 222
Gornick, Vivian, 36
Gould, Steven, 17, 22, 23, 222
Government, Australia, 164
Government, Federal, 98
Government, Indiana, 44
Government, Pennsylvania, 111, 138, 140, 215
Graham, Ronald, 95
Grahame, Kenneth, 129, 223
Graves, Robert, 30, 223
Green, Celia, 33
Greener, Leslie, 6, 88, 207
Groening, Matt, 8, 203, 210, 223, 229
Grunbaum, Branko, 65
Guardia, Fiorello La, 50, 225
Guin, Ursula Le, 119, 157, 225
Guiterman, Arthur, 27

H

Haida Indians, 122, 223
Haldane, John, 3, 15, 223
Halmos, Paul, 34, 68

Hamilton, W., 23, 223
Hammarskjold, Dag, 151, 223
Hamming, R., 64
Hamming, Richard, 64
Hardy, Godfrey, 49, 66, 73, 223
Harkness, Richard, 188
Harlow, Harry, 22, 223
Harrison, Al, 223
Harvey, William, 17, 20, 223
Hawking, Stephen, 4, 5, 36, 173, 223
Hawthorne, Nathaniel, 90, 223
Hay, John, 174, 223
Heaviside, Oliver, 36, 223
Heinlein, Robert, 20, 32, 39, 40, 56, 58, 70, 76, 78, 134, 177, 178, 181, 192, 195, 223
Heisenberg, Werner, 104, 178, 223
Heller, Joseph, 204, 223
Hemingway, Ernest, 201, 223
Hempel, Carl, 65
Henderson, F., 153
Henry, William, 171, 223
Heraclitus, 116, 153, 182, 223
Herodotus, 41, 156, 223
Hertz, Heinrich, 46, 223
Hilbert, David, 7, 8, 48, 172, 223
Hill, L. Erskine, 15, 223
Hinde, Robert, 41, 223
Hirst, Thomas, 71
Hitler, Adolph, 87, 223
Hoffer, Eric, 167, 223
Hofstadter, Douglas, 192
Holmes, Oliver, 118, 158, 223
Hooper, Grace, 34
Horace, 136, 223
Horton, Doug, 196
Howard, O., 128
Hoyle, Fred, 89, 223
Hubbard, Kin, 203
Hudson, William, 80, 224
Hugo, Victor, 39, 182, 224
Humboldt, Alexander von, 2, 52, 85, 86, 231
Hurston, Zora, 31, 224
Husson, Charlie, 170, 196, 224
Huxley, Aldous, 38, 62, 144, 196, 224
Huxley, Thomas, 3, 47, 52, 57, 224
Huygens, Christiaan, 63, 224
Huysmans, Joris, 114, 224
Hynes, H., 152

I

Iacocca, Lee, 177, 224
Ingalls, John, 147, 224
Institution of Civil Engineers, 111
Ives, Joseph, 108

J

Jackson, Andrew, 129, 224
Jacobi, Carl, 57, 224
James, William, 192, 224
Jeans, James, 14, 224
Jefferson, Thomas, 3, 62, 73, 101, 102, 133, 183, 191, 202, 224
Jeremiah, 126, 224
Job, 138, 174, 224
John Paul II, Pope, 142, 224
Johnson, Lyndon, 135, 224
Johnson, Samuel, 182, 197, 224
Johnstown Flood, 112
Jowett, Benjamin, 35, 224
Juvenal, 59, 224

K

Kadanoff, Leo, 55, 224
Kadlec, R., 151
Kandel, Stephen, 185, 224
Kant, Immanuel, 32, 54, 224
Kaplan, Abraham, 68
Karlin, Samuel, 52, 224
Kármán, Theodore von, 76, 231
Karr, J., 106, 153, 154
Keats, John, 137, 224
Keller, Helen, 70, 224
Kelley, John, 69
Kelvin, William, 6, 81, 224
Kennedy, John, 84, 204, 224
Kepler, Johannes, 53, 145, 224
Kettering, Charles, 188, 224
Keynes, John, 117, 224
Khaldun, Ibn, 67, 224
Khan, Faziur, 102
Khayyám, Omar, 197
Kilmer, Joyce, 151, 224
Kington, Miles, 147
Kipling, Rudyard, 99, 200, 225
Kissinger, Henry, 183, 225
Kleinhenz, Robert, 35
Kline, Morris, 48, 50, 183
Knight, R., 151
Koestler, Arthur, 4, 225
Koran, 10, 12, 130, 156, 191, 198, 225
Kosko, Bart, 50
Kronecker, Leopold, 69
Krutch, Joseph, 118, 225
Kuhn, Thomas, 117, 198, 225

L

La Guardia, Fiorello, 50, 225
Lagrange, Joseph Louis, Comte de, 15, 38, 189, 225

Lao-Tzu, 95, 131, 181, 225
Larcom, Lucy, 136, 225
Laurence, William, 93
Law, William, 81
Lawrence, D.H., 15, 21, 225
Le Guin, Ursula, 119, 157, 225
Leacock, Stephen, 6, 184, 225
Leach, Edmund, 30
Leahy, William, 92, 225
Leatherwood, Stephen, 160, 225
Lebesgue, Henri, 67, 225
Leeuwenhoek, Anton van, 18, 232
Leffler, Gosta, 66
Lehman, Arnold, 39, 225
Leibnitz, Gottfried, 45, 72, 94, 225
Lennon, John, 210, 225
Leopold, Aldo, 132, 134, 136, 140, 151, 167, 225
Leopold, Luna, 129, 225
Lewis, H.W., 80
Li Bings, 110
Li Erlang Temple, 99
Libchaber, Albert, 2, 64, 225
Lichtenberg, George, 37, 65, 225
Lien, Jon, 163, 225
Lilly, John, 161, 162, 225
Lincoln, John, 189
Lindbergh, Charles, 83, 117, 225
Linnaeus, Carolus, 16, 147, 225
Lipe, Chris, 12
Lippman, Gabriel, 49, 225
Littlewood, J., 47, 74, 177, 184, 226
Locke, John, 47, 176, 226
London, Jack, 202, 226
Loos, Anita, 14, 226
Lord Byron, 9, 222
Lorenz, Konrad, 48, 226
Lorimer, George, 211
Lovejoy, Thomas, 123, 226
Lovell, Jim, 84
Lucretius, 128, 226
Luke, 146, 226
Luther, Martin, 46, 226
Lytton, Edward, 226

M

Maccabees, 178, 226
Mach, Ernst, 45, 226
MacLeish, Archibald, 84, 226
Maidment, David, 132, 226
Mann, Thomas, 46, 226
Marinelli, Janet, 135
Markham, Edwin, 210, 226
Marx, Karl, 59, 189, 226
Masingill, M. Valere, 12
Mathesis, Adrian, 68, 70

Matthew, 32, 100, 150, 204, 226
Matthias, Bernd, 8
Maxwell, James, 6, 173, 177, 226
Mays, Larry, 132, 226
McClary, Michael, 108
McDonald, Mac, 194
McGlincy, Jim, 127, 226
McGrath, James, 28
McLuhan, Marshall, 91, 226
Mead, Margaret, 30, 226
Meadows, D., 123
Melville, Herman, 183, 226
Menaechmus, 71, 226
Mickey Mouse, 47, 226
Mill, John Stuart, 78
Mille, Cecil De, 211, 220
Millikan, Robert, 10, 92, 117, 180, 226
Milton, John, 50, 226
Mirsky, Steve, 79
MIT, 115
Mitchell, Brigadier General Billy, 83
Mitchell, Margaret, 71, 226
Mitchell, Maria, 2, 226
Mittermeir, Russell, 132, 150, 226
Monet, Claude, 148, 226
Monod, Jacques, 27, 226
Montesquieu, Charles, 72, 226
Moore, Ronald, 115, 227
Moran, Jim, 210
Mumford, Lewis, 97, 227
Munro, Hector, 175, 227
Murray, William, 128

N

NASA, 10, 119, 213, 219, 222, 224, 227
National Environmental Policy Act , 137, 227
National Resources Planning Board, 106
Navajo Indians, 145, 227
Neaves, Charles, 26, 227
Nebeuts, Kim, 69, 177, 91
NEPA, 71, 227
Neumann, Franz, 33
Neumann, John von, 52
Nevill, Dorothy, 198
New York Times, 87
Newell, Homer, 88, 227
Newman, James, 59, 64
Newman, Murray, 161
Newton, Isaac, 5, 44, 77, 105, 112, 113, 172, 227
Nicely, Thomas, 73
Nietzsche, Friedrich, 102, 19, 203, 205, 227
Nightingale, Florence, 51, 227
Nimoy, Leonard, 163, 227
Noble, Charles, 60
Norris, Ken, 160, 161, 163, 227
Noyes, H. Pierre, 7, 227

O

O'Connor, Flannery, 184, 227
Oglesby, Richard, 156
Ogyu, Sorai, 69, 227
Olla, Bori, 160, 227
Olson, Ken, 96, 227
Oppenheimer, Frank, 177
Orleans, Duchess of, 72
Ortega y Gasset, Jose, 133, 140, 227
Orwell, George, 164, 227
Osborn, Fairfield, 139, 227
Osler, Sir William, 175, 227
Ou-Tse, 34

P

Paracelsus, 13, 227
Pareto, Vilfredo, 192, 227
Parnas, Dave, 54
Pascal, Blaise, 86, 97, 173, 178, 182, 195, 196, 227
Paschkis, Victor, 80
Pasteur, Louis, 58, 191, 227
Patton, George, 190, 227
Paul, Epistles of, 198, 227
Paul, Pope John, II, 142
Pauling, Linus, 26, 227
Pearce, Donn, 198
Pearson, Karl, 51, 227
Peccei, Aurelio, 141
Peiper, Joachim, 112, 227
Peter, Irene, 198
Peters, Max, 80
Peters, Mike, 9, 227
Philpots, Edward, 129
Pierce, Benjamin, 62, 227
Pillsbury, Arthur, 109, 227
Pirsig, Robert, 57
Pius XII, 81, 228
Planck, Max, 117, 228
Plato, 45, 67, 68, 85, 164, 179, 181, 202, 228
Plautus, Titus Maccius, 81
Plotinus, 146, 228
Plutarch, 72, 228
Pochet, Pierre, 20, 228
Poff, N., 106, 153, 154
Poincare, Henri, 25, 31, 33, 65, 67, 228
Poisson, Simeon, 69, 228
Polya, George, 45, 46, 48, 228
Popper, Karl, 58, 228
Pordage, Matthew, 63
Porter, George, 179, 228
Potok, Chaim, 189, 228
Powell, Colin, 189, 193, 196, 199, 201, 203, 206, 228
Powers, Joan, 155
Powers, John, 185

Preece, William, 118, 228
Prestegaard, K., 106, 153, 154
Prinam, Stuart, 123, 228
Proverb, American, 80, 129, 130, 131, 132, 137, 149, 152, 155, 159, 164, 183, 194, 202, 205, 208
Proverb, Arabian, 204, 208
Proverb, Babylonian, 107
Proverb, Birdwatcher's, 165
Proverb, Chinese, 148–150, 209
Proverb, Ethiopian, 78
Proverb, Gardening, 149
Proverb, German, 163
Proverb, Greek, 128
Proverb, Indian, 44
Proverb, Jewish, 35
Proverb, Pilot's, 83
Proverb, Portugese, 131
Proverb, Russian, 130, 150, 197, 204
Proverb, Stock market, 208
Proverb, Weather, 130
Proverb, West African, 209
Proverbs, 105, 179, 188, 198, 228
Pryor, Karen, 161, 228
Psalms, 81, 156, 157, 166, 197, 228
Purcell, Edward, 45, 228

Q

Quetelet, Adolphe, 56, 228

R

Rabeni, Charles, 136
Raffel, Burton, 157
Rand, Ayn, 47, 228
Randers, J., 123
Reagan, Ronald, 142, 228
Redish, Edward, 53, 175, 175, 179, 180, 181, 209, 228
Reeves, Randall, 160, 228
Reilly, Victor, 211
Reisner, Marc, 109, 111
Renan, Ernest, 117, 228
Renyi, Alfred, 70
Reybold, Eugen, 103, 112
Rice, Grantland, 155, 228
Rice, Prudence, 29
Richards, Chet, 194
Richardson, Lewis, 104, 228
Richter, B., 106, 153, 154
Robinson, Edwin, 159, 228
Robinson, Michael, 80
Rockefeller, Laurance, 133, 228
Roddenberry, Gene, 86, 132, 228
Roentgen, Wilhelm, 33, 228
Rogers, Will, 101, 228

Romer, Robert, 174, 228
Rooney, Andy, 94, 228
Roosevelt, Eleanor, 137, 194, 228
Roosevelt, Franklin, 98, 182, 228
Roosevelt, Theodore, 109, 123, 133, 134, 191, 197, 228
Rosen, Barbara, 149
Rosenberg, Joel, 210
Rosenblueth, A., 53
Rosenlicht, Max, 68
Rota, Gain-Carlo, 47
Rousseau, Jean-Jacques, 78, 229
Rowland, F., 126, 229
Ruckelshaus, William, 142, 229
Russell, Bertrand, 35, 50, 58, 63, 64, 66, 116, 117, 206, 210, 229
Rutherford, Ernest, 6, 40, 49, 229

S

Sabaroff, Robert, 167
Safire, William, 87, 229
Sagan, Carl, 29, 33, 38, 114, 162, 229
Sakharov, Andrei, 93, 229
Sanskrit, 129
Sapolsky, Robert, 18, 229
Sappho, 29, 175, 229
Sarnoff, David, 207, 229
Saul, John, 175, 192, 198, 229
Scammon, Charles, 160, 229
Schaller, George, 22, 229
Schoklitsch, Armin, 215
Scott, Phil, 10
Seaborg, Glenn, 93, 229
Seinfeld, Jerry, 99, 229
Selous, E., 21
Selye, Hans, 18, 137, 229
Seneca, 146, 229
Seuss, Dr., 126, 229
Shakespeare, William, 46, 48, 76, 95, 144, 145, 193, 203, 229
Shaw, George, 12, 229
Shedd, John, 202
Shelley, Percy, 102, 229
Shephard, G., 65
Shiras, George, 125
Shoemaker, Eugene, 88, 229
Shoklitsch, Armin, 106, 111, 131, 229
Shozo, Tanka, 153
Sikorsky, Igor, 83, 229
Silverstein, Ken, 14, 93
Simmons, G., 62
Simon, Anne, 158
Simplicius, 11, 229
Simpsons, The, 8, 203, 210, 223, 229
Sinsheimer, Robert, 28
Smith, Clement, 103

Smith, Fred, 185
Smith, Hamilton, 28
Smith, Henry, 62, 229
Smith, Sydney, 62, 229
Smollen, Bill, 85, 88
Snepscheut, Jan van de, 57
Socrates, 113, 147, 229
Solomon, King, 200, 208, 229
Solter, Davor, 28
Sontag, Susan, 31, 230
Sophocles, ii, 230
Southwell, T., 161
Sowa, Scott, 136
Sowell, Thomas, 192
Sparks, R., 106, 153, 154
Speed, Adolphus, 163
Spencer, Herbert, 24, 230
Spender, Stephen, 50, 230
St. Paul's Church, 209, 230
Standen, Anthony, 15
Star Trek, 86, 94, 115, 132, 163, 164, 166, 167, 183, 185, 191, 194, 207, 214, 227, 230
Star Trek, The Next Generation, 115, 167, 230
Statius, Caecilius, 58, 230
Stein, Gertrude, 103, 230
Steinman, David, 139, 230
Steinmetz, Charles, 115, 230
Stevens, Charles, 109, 230
Stevens, Wallace, 52, 230
Stevenson, Adlai, 93, 230
Stevenson, Robert, 97, 230
Stewart, Potter, 137, 230
Sting, 141, 230
Stolzenburg, William, 167, 230
Stoppard, Tom, 51, 230
Stratton, James, 101
Straus, Michael, 101, 230
Strauss, Lewis, 93, 230
Streatfield, Geoffrey, 50, 230
Stromberg, J., 106, 153, 154
Strunk, William, 171
Strunsky, Simeon, 49, 230
Sumner, Gordon, 141, 230
Swain, Roger, 148
Swamp Land Act, 151, 230
Szent-Gyorgyi, Albert, 31, 230

T

Taft, William, 101, 230
Tagore, Rabindranath, 150, 230
Teller, Edward, 116, 230
Tesla, Nikola, 79, 230
Thales, 170, 230
Thatcher, Margaret, 200, 230
Thaves, Bob, 8, 97, 230
Theognis, 204, 231

Theoritus, 164
This is Spinal Tap, 209, 231
Thomas, David, 32, 231
Thomas, Lewis, 17–19, 25, 27, 28, 144, 288, 195, 231
Thompson, D'Arcy Wentworth, 19
Thoreau, Henry, 78, 103, 127, 147, 149, 150, 152, 154, 156, 165, 172, 189, 193, 200, 231
Thurber, James, 91, 231
Titchmarsh, E., 46
Titov, Gherman, 85, 231
Todhunter, Isaac, 73, 231
Tolstoy, Leo, 205, 231
Trevelyan, George, 182, 231
Trotsky, Leon, 196, 231
Truman, Harry, 92, 135, 231
Tsiolkovsky, Konstantin, 84, 119, 231
Turner, R., 162, 231
Turning, Alan, 95, 231
Twain, Mark, 25, 50, 55, 56, 59, 61, 107, 112, 162, 174, 182, 191, 194, 195, 196, 199, 208, 210, 231
Tyndall, John, 58, 231
Tzu, Lao, 95, 131, 181, 231
Tzu, Sun, 44

U

U.S. Army, 130, 156, 191
Udall, Stewart, 142, 231
Ulam, Stanislaw, 69
Updike, John, 11, 58, 231

V

van de Snepscheut, Jan, 57
van Gogh, Vincent, 145, 231
van Leeuwenhoek, Anton, 18, 231
Varberg, D., 26, 45
Veblen, Thorstein, 32, 231
Vest, Charles, 115, 231
Vinci, Leonardo da, (see da Vinci, Leonardo)
Virgil, 159, 231
Virginia Wetland Act, 152, 231
Vizinczey, Stephen, 133
Voltaire, 40, 231
von Braun, Wernher, 31, 85-88, 94, 192, 231
von Humboldt, Alexander, 2, 52, 85, 86, 231
von Kármán, Theodore, 76, 231
von Neumann, Johann, 45, 52, 96, 231

W

Walesh, 94
Walker, Frances, 51, 231
Walton, Izaak, 152, 232

Ware, Eugene, 173
Warner, Sylvia, 72, 232
Washington, George, 10, 152, 191, 232
Watson, James, 37, 205, 232
Watson, Lyall, 18, 232
Watson, Thomas, 96, 205, 232
Weber, David, 173
Weber, Robert, 15
Weil, Simone, 174, 190, 232
Weisbach, Julius, 10, 232
Weisskopf, Victor, 37
Welch, William, 34, 232
Wellington, Arthur, 79, 232
Wells, H.G., 16, 232
Westcott, Edward, 195
Weyl, Hermann, 59
Whistler, James, 146, 232
White, E.B., 133, 134, 149, 232
White, Gilbert, 154
Whitehead, Alfred, 3, 58, 119, 184, 192, 203, 232
Whitman, Christine, 124, 232
Whitman, Walt, 150, 232
Whitmann, Alden, 140
Wiener, Norbert, 38, 117, 178, 232
Wilbur, Carey, 191
Wilcock, Peter, 174, 232
Wilcox, Ella, 148
Williams, H., 203
Williams, Tennessee, 9, 208, 232
Wilson, Edward, 3, 17, 19, 23, 26, 40, 122, 126, 141, 142, 166, 167, 232
Wilson, Woodrow, 155, 232
Winn, Howard, 160, 232
Wittgenstein, Ludwig, 53, 232
Wolfe, Tom, 201, 232
Wordsworth, William, 144, 232
Wright, Frank Lloyd, 155, 232
Wright, Orville, 82, 232
Wright, Stephen, 86
Wright, Wilbur, 82, 232
Wulf, Bill, 76

X

X, Malcolm, 70, 232
Xenophanes, 155, 232

Y

Yablokov, Alexey, 160
Yevtushenko, Yevgeny, 89, 232
Yu, Emperor of China, 153, 232

Z

Zeuxis, 197, 232
Zwanzig, Carl, 79

About the Editors

Jon Fripp, (1967-), is a registered professional civil engineer and has undergraduate and graduate degrees in Civil Engineering from Virginia Tech. He has experience in the planning, analysis and design of hydraulic structures, in environmental restoration, and in working within large bureaucracies.

Michael Fripp, (1970-), has an undergraduate degree from Virginia Tech's Engineering Science and Mechanics Department and a graduate degree from M.I.T.'s Department of Aeronautical and Astronautical Engineering. His specialties are structural and acoustic dynamics, teaching, and viewgraph engineering.

Deborah Fripp, (1970-), has an undergraduate degree in biology from Stanford University and a graduate degree in marine biology from the Woods Hole Oceanographic Institution and from M.I.T. She has experience with dolphin behavior, stress hormones in rats, and governmental funding agencies.

I hate quotations. Tell me what you know.

— Ralph Waldo Emerson
in his journal, May 1849

More Great Books from LLH Technology Publishing

Digital Frequency Synthesis Demystified
by Bar-Giora Goldberg
INCLUDES WINDOWS 95/98 CD-ROM. An essential reference for electronics engineers covering direct digital synthesis (DDS) and PLL frequency synthesis. The accompanying CD-ROM contains useful design tools and examples, and a DDS tutorial by Analog Devices.
1-878707-47-7 $49.95

Video Demystified, Second Edition
A Handbook for the Digital Engineer
by Keith Jack
INCLUDES WINDOWS/MAC CD-ROM. Completely updated new edition of the "bible" for digital video engineers and programmers. Over 800 pages of hard-to-find design info and video standard specifications. The CD-ROM contains valuable test files for video hardware and software designers.
1-878707-23-X $59.95

Digital Signal Processing Demystified
by James D. Broesch
INCLUDES WINDOWS 95/98 CD-ROM. A readable and practical introduction to the fundamentals of digital signal processing, including the design of digital filters. The interactive CD-ROM contains a powerful suite of experimental, educational, and design tools. A volume in the Engineering Mentor series.
1-878707-16-7 $49.95

Programming Microcontrollers in C
by Ted Van Sickle
INCLUDES WINDOWS 95/98 CD-ROM.Shows how to fully utilize the C language to exploit the power of the new generation of microcontrollers that doesn't have to be programmed in assembly language. Also contains a great C tutorial for those who need it. Many practical design examples in over 400 pages.
1-878707-14-0 $39.95

Modeling Engineering Systems
PC-Based Techniques and Design Tools
by Jack W. Lewis
INCLUDES WINDOWS 95/98 CD-ROM.Teaches the fundamentals of math modeling and shows how to simulate any engineering system using a PC spreadsheet. A great hand-holding introduction to automatic control systems, with lots of illustrations and practical design examples. A volume in the Engineering Mentor series.
1-878707-08-6 $29.95

Controlling the World with Your PC
by Paul Bergsman
INCLUDES PC DISK. A wealth of circuits and programs that you can use to control the world! Connect to the parallel printer port of your PC and monitor fluid levels, control stepper motors, turn appliances on and off, and much more. The accompanying disk for PCs contains all the software files in ready-to-use form. All schematics have been fully tested. Great for students, scientists, hobbyists.
1-878707-15-9 $35.00

Bebop to the Boolean Boogie
An Unconventional Guide to Electronics Fundamentals, Components, and Processes
by Clive "Max" Maxfield
The essential reference on modern electronics, published to rave reviews from engineers, educators, and nontechnical types who need to work with technology. Covers all the basics from analog to digital, bits to bytes, to the latest advanced technologies. 500 pages of essential information presented with wit and style. Worth the price for the glossary alone!
1-878707-22-1 $35.00

Fibre Channel, Second Edition
Connection to the Future
by the Fibre Channel Industry Association
A concise guide to the fundamentals of the popular ANSI Fibre Channel standard for high-speed computer interconnection. If wading through the entire Fibre Channel standard document seems too daunting, this is the book for you! It explains the applications, structure, features, and terminology in plain English.
1-878707-45-0 $16.95

The Forrest Mims Engineer's Notebook
by Forrest Mims III
A revised edition of a classic by the world's bestselling electronics author. Includes hundreds of circuits built from integrated circuits and other parts available from convenient sources. Also contains special tips on troubleshooting, circuit construction, and modifications.
1-878707-03-5 $19.95

Simple, Low-Cost Electronics Projects
by Fred Blechman
Whether you're a beginner or an old hand, let popular author Fred Blechman be your guide to the exciting world of electronics. Begin with simple designs and progress to more sophisticated projects, using commonly available parts. You're guaranteed success with this essential book on your workbench!
1-878707-46-9 $19.95

The Integrated Circuit Hobbyist's Handbook
by Thomas R. Powers
This practical circuit collection belongs on every electronics hobbyist's shelf! Covers the major types of ICs and provides complete detail and theory about their operation. Also includes directions for building electronic devices that make use of ICs, as well as a massive listing of the most popular ICs in the world, thoroughly indexed by application.
1-878707-12-4 $19.95

The Art of Science: A Practical Guide to Experiments, Observations, and Handling Data
by Joseph J. Carr
A friendly and readable guide to the "nuts and bolts" of scientific inquiry. Examines the scientific process in detail, covering experimental technique, error analysis, statistics, graphing, and much more. A great reference for students, engineers, scientists.
1-878707-05-1 $19.95